State, Society, and Tribes: Issues in Post-Colonial India

State, Society, and Tribes: Issues in Post-Colonial India

Virginius Xaxa

Pearson

ISBN 978-93-325-3459-9

First Impression, 2014

Published by Pearson India Education Services Pvt. Ltd, CIN: U72200TN2005PTC057128.

Head Office: 15th Floor, Tower-B, World Trade Tower, Plot No. 1, Block-C, Sector 16, Noida 201 301, Uttar Pradesh, Indi
Registered Office: 7th Floor, SDB2, ODC 7, 8 & 9, Survey No. 01 ELCOT IT/ ITES-SEZ, Sholinganallur,
Chennai – 600 119, Tamilnadu, India.
Phone: 044-66540100
Website: in.pearson.com, Email: companysecretary.india@pearson.com

Printer: Manipal Technologies Limited, Manipal

To my wife Chameli

Contents

Preface and Acknowledgements

My interest in the studies of tribes emerged more from my association with the Department of Sociology at the Delhi School of Economics and its intellectual and social environment than out of my own previous research interests, which were in the field of agrarian and development studies. My intellectual engagement with the tribal question in India was generated by conversations and discussions with colleagues on issues and problems concerning students belonging to the scheduled castes and scheduled tribes in the department and the university. This volume is an outcome of that engagement. Needless to say, I owe an intellectual debt to my colleagues in the department. I am particularly grateful to Professor Veena Das for inviting me to join *The Oxford Indian Companion to Sociology and Social Anthropology* placing before me a field that would eventually offer me an intellectual journey. I am also grateful to Professors Andre Béteille and J. P. S. Uberoi for their keen interest in my work. Interacting and engaging with them has been a great learning experience and a source of much intellectual enrichment. I would like to acknowledge my gratitude to Professor Anand Chakravarty for his faith and confidence in me and to Professor Rajni Palriwala and Dr Rita Brara for reading and commenting on some of the chapters.

Outside the department, I am most grateful to Professor Sharit Bhowmik, Dr Ritambhara Hebbar, Dr P. D. Khera and Dr Ashley Tellis. Professor Bhowmik, formerly at the Department of Sociology, Delhi School of Economics and now at TISS, Mumbai, has been a source of great support all through my career. Ritambhara Hebbar, a faculty member at TISS and an ex-student of mine read many of the papers included in this volume and provided useful comments and suggestions. Dr P. D. Khera shared with me his rich and invaluable ideas and insights in many conversations with him on the tribal question in India. Dr Tellis was generous in helping me read the proof. I am grateful to Pearson Education India and especially to Debjani M. Dutta for her keen interest in the publication of this volume.

This endeavour would not have come to fruition but for the support that I have received from my family. I am deeply indebted to my son Aashish, who, despite my considerable negligence toward him as a father, has been more than patient with me. Chameli, my wife, has been a constant source of support. In spite of her ill-health, she took upon herself many of the responsibilities that were mine to enable me to carry on with this work. My indebtedness to her could not be expressed in words. I dedicate this volume to her.

I wish to thank the following for permission to reproduce the articles in this volume:

Economic and Political Weekly for 'Transformation of Tribes in India: Terms of Discourse' XXXIV (24), 1999; 'Tribes as Indigenous People of India' XXXIV (51), 1999; 'Protective Discrimination: Why Scheduled Tribes Lag Behind Scheduled Castes' XXXVI (29), 2001; and 'Politics of Language, Religion and Identity' XL (13), 2004.

Indian Journal of Gender Studies for 'Women and Gender in the Study of Tribes in India', 11:3 (2004).

Sage Publications for 'Empowerment of Tribes' in Debal K. Singharoy (ed.), *Social Development and the Empowerment of Marginalized* (New Delhi: 2001).

Rawat Publications for 'Reading Tribes in the Light of Citizenship Rights' in M.N. Karna (ed.), *Democracy, Pluralism and Conflict* (Jaipur: 2005).

Chapter 1

Introduction

Differences have been the hallmark of Indian society. These differences emerged as the result of various long and complex historical processes. The marks of differences have been diverse and varied, but the major ones have been located primarily around those of caste, language, religion, and region. To these principal marks of difference, the new one of tribe was added during the colonial period. Before the colonial period, the use of a generic term to describe tribal peoples was not in existence.

Some anthropologists have described the phenomena of caste and tribe as colonial constructions in the sense that the character of these groups was solidified by the British through processes of classification and enumeration. Béteille argues that this is truer of tribes than it is of castes. He holds that there was already a complex and sophisticated literature on the caste system before the adoption of British initiatives. Further, the use of the category of caste was not confined to the intelligentsia only but was also a part of the thinking of the common man much before the advent of the British. However, hardly anything corresponding to this existed in the case of those we know as tribes today. It is on this count that the category of tribe has been talked about as a colonial construction (Singh 1993; Béteille 1993). This statement is not meant to convey the claim that groups so identified did not have a distinct identity of their own. Indeed, not only did these groups see and identify themselves as being different, they were also seen as being different by others. However, the general category of tribe was absent. There were also no tribal writers or scholars to reflect on and write about the nature of tribes. Discussions of the nature and significance of tribe as a distinct category had to wait until the coming of colonial rule. The consciousness of the distinct and separate identity of all the tribes in India taken as a whole is a part of modern consciousness, brought into being by the colonial state and confirmed by its successor after independence. Ray does talk of a category of people in Indian history now generally identified as tribes. These were the janas with an egalitarian form of social organization as against those with a jati system of social organization (Ray 1972). Other scholars caution against such a view. They point out that any attempt to identify jana with present-day tribals is not without difficulty. They say that the category is not only amorphous but also overlaps with that of non-tribe. Béteille argues that the distinction between jana and jati must have been less clear in ancient times than the corresponding distinction between tribe and caste today (Béteille 1986; Singh 1993; Roy Burman 1994).

Colonial administrators used the term tribe to describe people who were heterogeneous in physical and linguistic traits, demographic size, ecological conditions of living, regions inhabited, stages of social formation, and levels of acculturation and development. The need for such a category was necessitated by a concern to subsume the enormous diversity into neat and meaningful categories for both classificatory purposes and administrative convenience. Hence, although tribe as a category and as a point of reference may be treated as a colonial construction, the image and meaning underlying the category was far from being a colonial construction. (Indeed, there was much similarity between the connotations of the term tribe and the ideas and images constructed by Indian civilization of groups and communities that came to be designated as tribes during the colonial period.) The term tribe since the sixteenth century has referred to groups and/or communities living under primitive and barbarous conditions. Sanskritic and Hindu religious texts and traditions describe and depicte tribes in a similar fashion. Bara takes the point even further when he states that the pre-colonial depiction of the tribal people of India as dasyus, daityas, rakshasas, and nishadas, when juxtaposed with mid-nineteenth-century Western racial concepts, advanced the aspect of bestiality associated with tribes (Bara 2002: 125).

The study of groups, which subsequently came to be described as tribal studies, began with the establishment of the Asiatic Society of Bengal in 1784. Since then, scholar– administrators have been writing general works on the people of different regions. These works were more in the nature of inventories and took the form of handbooks and monographs. Notable among these works, focusing on groups and communities that later came to be called tribes, are Dalton's *Descriptive Ethnology of Bengal* (1872), Risley's *Tribes and Castes of Bengal* (1891), Russell and Hira Lal's *Tribes and Castes of the Central Provinces of India* (1916), and Thurston's *Castes and Tribes of Southern India* (1909). Vidyarthi describes this phase of studies on the tribes of India as the formative period, dating from 1784 to 1919. He identifies and describes the next two phases as constructive (1920–49) and analytical (1950 onwards). The constructive phase is the period when anthropology was introduced into university curricula in India, and the analytical phase saw the adoption of analytical and action-oriented approaches in the study of tribes (Vidyarthi 1982). There has been a flood of literature on tribes in the post-independence period. These have come from mainly three sources: the Anthropological Survey of India (ASI), university departments of sociology and anthropology, and tribal research institutes set up in states having a substantial tribal population. The tribal research institutes were established in the mid-1950s mainly to undertake problem-oriented research studies for the effective formulation and implementation of developmental programmes in tribal areas.

When the British began to write on India in the eighteenth century, they made no distinction between tribe and caste. Rather, the two terms were used synonymously or even cognately, notwithstanding the fact that the British did treat a segment of the population known later as tribes as different from the rest of the Indian population. The difference was conceived more in terms of ethnicity than in terms of caste and tribe. This can be inferred from the nature of the administrative systems in vogue in tribal and non-tribal areas. The two sets of population had by and large different administrative set-ups. Laws meant for the general population were usually not applicable in the case of groups called tribes. More often than not, special laws, that is, laws in consonance with the tribal system of administration, were framed for their governance. In colonial administrative parlance, the nature of such administration was described by terms such as non-regulation tracts, scheduled areas, or excluded and partially excluded areas. The difference was invariably linked with different conceptions of the two sets of people.

The attempt at delineating tribes began with the need to provide detailed and classified information about people in the census. Accordingly, a certain group of people was categorized as a tribe when the census began to be undertaken in the late nineteenth century, though the criteria used were far from unambiguous. After 1901, however, a somewhat clearer criterion began to be used. Tribes were identified and described as those groups that practiced animism; later the phrase tribal religion was used in its place. The use of this criterion was continued in subsequent census enumerations, but some other dimensions were also added. Thus, the 1921 census report described them as hill and forest tribes, and in 1931, when Hutton was the Census Commissioner, tribes were also referred to as primitive tribes. Through the adoption of descriptions such as hill and forest tribes and primitive and backward tribes, the elements of geographical isolation and primitive living conditions were added to the distinction between tribes and non-tribes. These elements were, however, never explicitly employed in delineating tribes in the census enumeration. What was employed explicitly was the aspect of religion. That is, tribes were those groups that did not adhere to religions such as Hinduism, Islam, and Christianity. If a group were shown to be Hindu in its beliefs and religious practices, it was identified as a caste. If it were shown to be animist, it was treated as a tribe. This distinction was not satisfactory for it is not difficult to show that caste Hindus practiced elements of animism.

Subsequently, however, a list of tribes began to be drawn up with a view to giving them certain political and administrative concessions. In doing so, the consistent application of the criteria was given a back seat. The list of scheduled tribes included in the Constitution had its genesis in these administrative and political considerations. It is obvious then that the list was drawn up more on the basis of administrative and political considerations, although it came to be accepted as part of scholarly vocabulary.

After independence, greater attention was paid to the criteria for determining tribes. However, even today, there is no agreement among scholars on this subject. The Tribal Welfare Committee of 1951, consisting of anthropologists, administrators, and social workers, set up under the auspices of the Indian Conference of Social Work in Calcutta, deliberated on the criteria but could not arrive at a clear set. In the early 1950s, in an attempt to adopt criteria for identifying scheduled tribes, the Commissioner for scheduled castes and scheduled tribes sought the views of different provinces and states regarding prominent characteristics that might distinguish these communities. The criteria put forward by the different provincial governments make for interesting reading. They include such features as physical characteristics, linguistic affiliation, cultural contact, occupation, and ecological considerations (Khubchandani 1992). In the context of tribes, which are seen as not yet contaminated by the influence of Hindu civilization, the criteria used to distinguish them include such elements as living in inaccessible places, speaking a tribal dialect, practising animism, and engaging in primitive occupations. In addition, such groups were said to belong to either Negrito, Australoid, or Mongoloid stock, with nomadic habits and a love of dance and music. Different and even contradictory criteria have been used by scholars, administrators, and reformers for the purpose of identifying tribal groups. Thus how to use the term tribe in the Indian context, and what we mean by tribal society, has not been an easy question to resolve.

Tribes in India are thus defined not so much in terms of *coherent and well-defined criteria* but in terms of the *administrative* classification that divides the population into tribal and non-tribal. Tribes are treated as those groups enumerated in the Indian Constitution in the list of scheduled tribes. Indeed, the Constitution defines a scheduled tribe 'as such tribe or tribal community or part of or groups within such tribes or tribal communities as are deemed under Article 342 to be scheduled tribes'. And since the list is associated with the extension of administrative and political concessions and benefits to the groups mentioned therein, there has been little attempt to relate it to the definition of tribe. This does not mean that the politico-administrative category has not been examined for its efficacy as a scientific tool. Indeed, where such an exercise has been undertaken, it has been shown to have little scientific use. In other words, the category of tribe as used in anthropological writings is shown to be valid for a very limited range of groups in the list. And yet the term tribe is not restricted to referring only to these groups. That there is a discrepancy between the concept and its application to different groups and communities listed as tribes is obvious. Yet there seems to be no sign of the term being abandoned. There is then a certain special sense in which the term tribe has come to acquire its meaning in the Indian context, which is quite different from the way in which it has been used in the anthropological literature. Béteille's offers an apt description of this situation. He writes that when tribes and civilization coexist as in India and the Islamic world, being a tribe has been more a matter of remaining outside the state and civilization, whether by choice or by necessity, than a matter of attaining a definite stage in the evolutionary advance, from the simple to the more complex. Therefore, we cannot dismiss as anomalous the Indian practice of regarding as tribe a large assortment of communities differing widely in size, mode of livelihood, and social organization. They are all tribes because they all stood more or less outside of Hindu civilization and not because they are at exactly the same stage of evolution (Béteille 1986).

Given that the question of tribes in India is closely linked with administrative and political considerations, there has been more concern with the identification of tribes than with their definition. This statement is not meant to convey that lists have been drawn up without any conception of tribe whatsoever. There did exist some conception. This was obvious from the criteria that were adopted. These included such features as geographical isolation, use of simple technology, living conditions, general backwardness, practice of animism, tribal language, and physical features. The problem, however, lay in the fact that the criteria were neither clearly formulated nor systematically applied. One set of criteria was used in one context, and quite another employed in another. The result is that the list includes groups and communities strikingly different from each other in respect of not only the size of population but also the level of development. Indian anthropologists are acutely aware of a certain lack of fit between what their discipline defines as tribe and what they are obliged to describe as tribe. Yet they have continued with the use of the term.

In view of the above, there has been a lot of dissatisfaction over the use of the term tribe among scholars, administrators, and social workers. Alternative terms have been adopted, such as aborigines/aboriginals (Risley 1903; Elwin 1944), 'so-called aborigines', backward Hindus (Ghurye 1963), ethnic minorities (Pathy 1988), the fourth world (Sengupta 1982), and tribes in transition (Desai 1960). However, these alternatives have failed to become general terms of discussion in the literature on tribes. New terms have been coined in recent years, such as the marginalized, dalits, and adivasis (indigenous peoples). The term 'the marginalized' is too general and is used to refer to wide segments of the population. Hence, it has not gained as wide a currency as the other two terms, namely dalit and adivasi. Activists and even social scientists widely employ the term dalit to refer to the former untouchables (scheduled castes) and tribes. Dalit in its English translation means oppressed people. In this sense, it is at times used to refer to wide sections of the population, such as peasants, workers, and women. More strictly, however, activists and social scientists, as noted earlier, use it to refer to the scheduled castes and scheduled tribes. It has emerged from a certain context and from the experience of a certain category of people, namely the former untouchables. The term dalit describes this consciousness and the people who have had this experience are called dalits. Since scheduled castes and scheduled tribes suffer problems of a similar nature, such as exclusion, deprivation, and discrimination, and since there are constitutional provisions and special policies and programmes for their protection and welfare, there has been a general tendency to place them under the general category of dalit. However, the use of the term dalit to describe tribal populations betrays the social consciousness of the scheduled castes. Although tribal people suffer from the same sorts of problems as the scheduled castes, their historical experiences and social consciousness are markedly different from those of the scheduled castes. Hence to extend the category of dalit to tribes is to impose an alien consciousness on them. The kind of consciousness that tribes have of their own selves is not one of dalits but that of adivasis. It is a label that others have given to them, but tribes over the course of time have internalized it. It is through this category that they think of and relate themselves to others. Since the UNO's declaration in 1993 of the International Decade of the Indigenous Peoples, the category adivasi has become a powerful point of reference for social and political mobilization in India. The usefulness and significance of the term indigenous peoples in the context of India will be critically examined later in this book.

Language and region, each on its own, are important marks of difference, but they tend to coincide with each other in the Indian context. In fact, the two together have been the driving force behind the reorganization of society and polity in India in the post-independence period. Region and language, though important, do not tell us anything about the inner workings of the society and polity that they embody. What is even more striking is that despite differences on the basis of language and region, the societies so marked are enormously similar in their societal characteristics—religion and caste being predominant among them. Indeed, religion and caste cut across regions and languages, and to that extent they constitute a common thread across the diversity of languages and regions (Dumont 1998). It is this commonality that has come to be the principal unit of sociological study and discourse in India. Again, it is against this commonality of Indian society that tribes in India have been posited and constituted by sociologists and social anthropologists.

Consequently, the study of tribes has primarily been conducted from the perspective of caste. Caste being a unique and pervasive feature of Indian society, such a perspective may have been inevitable. Hence tribal studies in India have become caste-centric. Tribes have been seen as the natural equivalent of caste; a contrast has been drawn between tribe and caste, and even change among tribes has primarily been studied in terms of their transformation into castes. Paradoxically, this has been the case despite the fact that tribes have been seen primarily not only as a society but also as a particular type of society. Ideally, then, the contrast should have been not with caste but with society, as in the case of Oriya, Bengali, Telugu, etc. since the tribes represent all those features that are characteristic of a society. The caste orientation in tribal studies was reinforced in the post-independence period. Constitutional provisions for scheduled tribes and scheduled castes have further strengthened the caste orientation and perspective adopted towards an understanding of tribal society. That is, tribes are invariably seen as caste-like entities rather than as distinct societies.

The So-called Tribal Policy

After India became independent, formulating and implementing an appropriate policy for tribes assumed critical importance. Even before independence, the nature of the policy to be adopted had become a matter of heated debate between two scholars, namely Elwin and Ghurye. The debate had its genesis in Elwin's *The Baiga*, published in 1939. The argument made in the book was forcefully carried forward in a pamphlet, 'The Aboriginals', published in 1944. Here, Elwin made certain suggestions regarding the approach to be adopted towards tribes. The issue at the centre of the debate was whether the current state or condition of isolation of the tribes should be continued or whether they should be drawn into the larger society, so as to enjoy the benefits available to the general population. Elwin was charged with advocating the policy of isolation. The charge, Elwin (1960b) wrote later, was a result of suggestions badly put and the unfortunate connotation of the expression 'national park' used by him. The suggestion, he said, was made in desperation as the Baigas had very little left of their own when the book was written in 1939. It was not so much a matter of preserving their culture as of keeping them alive and protecting them from exploitation and oppression. Since then, there was a shift in Elwin's position and he no longer advocated the policy of isolation. One finds on closer examination that the policy was meant for a very small section of the tribal population. The policy Elwin advocated for the majority of the tribal population was no different from the policy advocated for the general population, for he believed that the problems faced by the tribal population could not be considered apart from those faced by the general village population. What Elwin advocated was not the same policy for the entire tribal population of the country (Elwin 1960b).

As against this, Ghurye advocated the policy of assimilation. He argued that the tribes were Hindus, or, to put it more bluntly, backward Hindus. He thought of them as inferior and backward because they were cut off from the large mass of the Indian population. He also identified them as standing a step lower on the evolutionary ladder, and in fact, not being very different from the mass of the Indian population. His considered view was that tribes should be assimilated into the larger society so that they could be lifted out of their abject poverty and base living conditions. This view placed tribals at the lowest position on the tribe–caste–class continuum, a conceptual paradigm used for understanding the processes of change in tribal life. For Ghurye, this continuum suggested a movement of the tribals from tradition to modernity, from pantheism to a higher form of religion, from a base and crude way of life to one morally and ethically superior. This was a way of tracing the assimilation–integration stage of the tribals into the national mainstream (Ghurye 1963).

Neither of the two policies was thought adequate in independent India by the nationalist leadership. This is evident from the nature of the provisions laid down in the Indian Constitution for tribals. The provisions pointed to an approach that was quite different from those propagated by Elwin and Ghurye. The provisions included statutory recognition, proportional representation in the legislatures, the right to use their own language for education and other purposes, the right to profess their own faith, and the right to pursue development according to their own genius. The Constitution also empowered the state to make provision for reservation in jobs and appointments in favour of tribal communities. In addition, the Directive Principles of State Policy of the Constitution required that the educational and economic interests of the weaker sections of society, including tribals, be especially promoted. Moreover, the Constitution empowered the state to bring areas inhabited by tribes under the Fifth and Sixth Schedules for the purpose of special treatment with respect to the administration of tribal people. These constitutional provisions thus aimed at safeguarding, protecting, and promoting the interest of tribal peoples. If one were to examine these provisions more carefully, one would find that the Constitution clearly adopts a policy of integration rather than of isolation or assimilation, albeit without using the term and concept of integration even once. Indeed, there is no official document where the policy of integration finds an explicit formulation. Yet this is what goes in the name of tribal policy in India, not without reason, of course, if one were to take note of the constitutional provisions. In fact, the five principles mentioned by Nehru in his foreword to Elwin's *A Philosophy for NEFA*, which was proclaimed as the guide for tribal development, were no more than an enunciation of the principles underlying the constitutional provisions. The principles as enunciated by Nehru were:

(1) People should develop along the lines of their own genius, and we should avoid imposing anything on them. We should try to encourage in every way their own traditional arts and culture.
(2) Tribal rights in land and forest should be respected.
(3) We should try to train and build up a team of their own people to do the work of administration and development. We should avoid introducing too many outsiders into tribal territory.
(4) We should not over-administer these areas or overwhelm them with a multiplicity of schemes. We should rather work through, and not in rivalry to, their own social and cultural institutions.
(5) We should judge results not by statistics of the amount of money spent, but by the quality of human character that is evolved (Elwin 1960a).

These principles have been the guidelines of tribal development more for the north-eastern region than for mainland India, where the reality has been a situation characterized more by assimilation than by integration.

The State Agenda for Tribals

The Constitution confers fundamental rights on all citizens irrespective of their caste, class, language, and creed. In constitutional and legal terms, tribes have also been given the same status as other citizens. This is an important step in building an inclusive society. In addition to the fundamental rights, the Constitution contains many special provisions for tribal people. These include provisions for their statutory recognition (Article 342); for their proportional representation in Parliament and the state legislatures (Articles 330 and 332); restrictions on the right of ordinary citizens to move and settle in tribal areas or to acquire property there (Article 19(5)); the protection of tribal language, dialect, and culture (Article 29); and for reservation in general (Article 14(4)) and in jobs and appointments in favour of tribal communities in particular (Article 16(4)). The Directive Principles of State Policy suggest that the educational and economic interests of the weaker sections of society, including tribes, be especially promoted (Article 46). In addition, the Constitution empowers the state to bring the areas inhabited predominantly by tribes under special treatment for administrative purposes. Tribes have thus not only been endowed with citizenship rights but also with certain special rights. It is against this backdrop that the state's agenda for tribes has to be understood. The different measures taken for their upliftment can be divided broadly into three categories, namely protective, mobilizational, and developmental.

Protective Safeguards

The Constitution aims to protect and safeguard the interests of tribal people. Keeping this in mind, laws have been enacted in almost all the states with tribal populations to prevent alienation of land from tribes to non-tribes. In some parts, such acts have been in existence since the British period. In the post-independence period, various laws were adopted for the restoration of tribal land. Protection is also provided in the form of special administration of tribal areas. In administrative parlance, such areas are referred to as the Fifth and Sixth Schedule areas (Articles 244 and 244[a]). The Fifth Schedule provides for special legislative powers for the governor, the governor's report, and, most importantly, the tribal advisory council. The Sixth Schedule provides for autonomous district councils, thus offering scope for tribal self-governance. The Sixth Schedule in general covers the north-eastern region; the Fifth Schedule is in operation in other regions where tribes form a majority in a district or districts.

Mobilizational Strategies

Mobilizational strategies refers to the reservation extended to tribals in fields such as education, employment, and politics. In each of these spheres, a certain percentage of seats are earmarked for the scheduled tribes. The quota was fixed keeping in mind the size of the scheduled tribe population when the measure was introduced. Accordingly, 7.5 per cent of seats were reserved for tribes in all these spheres. The size so fixed has remained the same until today even though the size of the population has changed. The provision of reservation in politics meant reservation of seats in Parliament and the state legislatures, meant to be in force for a period of 10 years. However, it has been renewed at the expiry of every 10-year period. It is important to note that such a time limit, contrary to general belief, was not fixed in the areas of government service and educational institutions. There were no reservation in the institutions of local self-governance until the 73rd Amendment to the Constitution.

Developmental Measures

Developmental measures include programmes and activities meant for the uplift and progress of tribal people. Efforts in this direction began with community development programmes, with the block as the main administrative unit. This was replaced by the tribal sub-plan approach under the Fifth Plan (1974–79). The approach has proved to be relatively effective in ensuring the flow of population-proportionate funds for tribals from the general development sector. Provisions have also been made for special central assistance to state tribal subplans. Poverty-alleviation programmes were launched under the Sixth Plan (1980–85). Under the Seventh Plan (1986–1990), two national-level institutions were set up, namely the Tribal Cooperative Marketing Development Federation (TRIFED) and the National Scheduled Castes and Scheduled Tribes Finance and Development Corporation (NSFDC). The aim of TRIFED is to pay remunerative prices for the forest and agricultural produce of tribes. The aim of NSFDC is to provide credit support for employment generation.

Despite the constitutional provisions and the special policies and programmes adopted by the state for the welfare and benefit of tribal people, the condition of the tribals continues to be deplorable. Land, the life-support system of tribals, continues to be passed from tribes to non-tribes through such means as fraudulent transfers, forcible eviction, mortgages, leases, and encroachments despite legislation restricting the alienation of lands from tribes to non-tribes. As per the information available with the Ministry of Rural Development in January 1999, 465,000 cases of alienation of tribal land, covering an area of 917,000 acres, were registered in Andhra Pradesh, Assam, Bihar, Gujarat, Himachal Pradesh, Karnataka, Madhya Pradesh, Maharashtra, Orissa, Rajasthan, and Tripura (Planning Commission 2001: 39). Displacement is the other important factor accounting for the loss of control and ownership over land and forest by the tribals. Displacement has occurred primarily because of large-scale projects like industries, irrigation and hydraulic projects, and mining. Projects such as these resulted in the displacement of a total of 21.3 million people in the period 1951–90 in Andhra Pradesh, Bihar,

Gujarat, Maharashtra, Madhya Pradesh, Rajasthan, and Orissa. Of this number, 8.54 million, that is, 40 per cent, were tribals (Planning Commission 2001: 39). The large-scale alienation of land from tribes to non-tribes and the massive displacement of tribal people largely explain why the percentage of tribal cultivators has declined from 68.15 in 1961 to 54.5 in 1991. Conversely, the percentage of agricultural labourers has increased from 19.71 in 1961 to 32.69 in 1991 (Planning Commission 2001: 91) Further, as much as 42.9 per cent of the operational holdings of tribes belongs to the category of marginal farmers, which means that they hold less than one hectare (Ministry of Tribal Affairs 2001: 36). It is, therefore, not surprising that a large proportion of the tribal population lives in poverty. In 1993–94, 51.14 per cent of tribal people lived below the poverty line compared to 35.97 per cent of the general population (Ministry of Tribal Affairs 2001: 43). About 80 per cent of tribal children are anaemic and 50 per cent of them are underweight (Planning Commission 2001: 23). The incidence of anaemia among tribal women is as high as 65 per cent compared to 48 per cent for the general population. Infant mortality and under-five mortality per thousand live births are as high as 83 and 126.6 respectively compared to 61.8 and 82.6 respectively for the general population (NFHS-II 2000). Even in the sphere of education, where tribes are said to have made some tangible progress, the scene is far from satisfactory. The literacy rate among tribes in 1991 was 29.6 per cent as compared with 52.2 per cent for the general population. The female literacy rate among tribes was 18.2 as against 39.3 per cent for the general population. What is a matter of great concern is that the gap between tribes and non-tribes has been widening steadily. The gap was 19.8 per cent in 1981 and increased to 22.6 per cent in 1991.

Through constitutional provisions and measures such as those described above, the state has attempted to encourage the inclusion of tribes in the larger society. Conferment of citizenship rights (civil and political rights) has been one of the major ways of building an inclusive society. To build this inclusive society not in the formal sense but in the substantive sense, special measures for the benefit and development of tribes and to increase their representation in state institutions were adopted. Attempts at building this inclusive society have been relatively successful in the direction of extending civil and political rights to tribes and in increasing their share and participation in state institutions such as Parliament and state legislatures, government services, and public sector undertakings. This aim has been to bring tribes closer to the state and the larger Indian society through measures such as reservation in politics, education, and government employment. However, the attempt was most unsuccessful precisely in the area meant to protect tribal culture and tradition, crucial to the ethos of integration advocated as the national tribal policy by scholars and administrators. Protection of tribal land has also been a failure, but at least there has been some tangible effort in this regard both in terms of legislation and implementation, even though the latter has been highly unsatisfactory.

This volume seeks to address the questions emanating from the state agenda for tribes in postcolonial India. After all, much of what has been witnessed among the tribes in postcolonial India has more to do with the state's role and policy than with other factors. In it, I seek to assess the impact of this agenda on tribes and the tribal peoples' responses to it. Both the effects and the responses are diverse and complex, and it is not possible to engage with them all here. Hence only a few of the most important issues are examined in a critical and reflective way.

The book comprises 10 chapters. The introductory chapter examines the evolution and use of the concept of tribe in India. The term was initially used to refer to a wide range of groups but later came to be restricted to specific groups. The shift was a result of greater clarity about the criterion or criteria that came to be associated with the concept. Nevertheless, the concept suffers from a certain inadequacy. This is most evident in its inability to grapple with the actual empirical social reality. The mismatch between the concept as it is understood and the groups and communities described in India as tribes is a reflection of this inadequacy. Hence there is a general uneasiness among scholars over the use of the term or category 'tribe' in India. The second part of the introduction deals with the question of state policy and agenda-making for tribals and outlines some of the critical issues emerging from these developments and initiatives. The critical issues dealt with here include land, forest, administration, reservation, language, and religion.

Since the changes among tribes go well beyond interventions by the state, whether colonial or postcolonial, Chapter Two discusses the nature of tribal transformation in India. The transformation of tribes in India has been studied primarily from three perspectives, namely caste, peasant, and social differentiation. Of these, the most pervasive and dominant has been the perspective on the transformation of tribe into caste. The chapter critically examines the existing literature. It problematizes the concept of Sanskritization, which scholars have used widely for understanding this transformation. The relative effectiveness of the terms and Hinduization and the place of language in the acculturation process have been subjected to scrutiny. The distinction between culture and structure, which has been generally overlooked in the discussion on tribal transformation, has been discussed. Through the critical deployment of these terms and a critical assessment of the processes involved, the assertion of the process of tribal transformation into caste is questioned and problematized.

There has been much dissatisfaction with the use of the term or category tribe. Scholars have considered alternative terms and categories, all of which have been found wanting in one sense or the other. One of the terms or categories that has received wide currency and recognition is adivasi, or indigenous peoples. Yet the use of this category is not without problems, at least in the Indian context. Chapter Three looks at the problems that arise from the employment of the category of indigenous peoples in the Indian context. It examines the arguments advanced in favour as well as against this usage. It looks at why the use of this category has become such a contentious issue in India in recent years, though the term in its Sanskritic version, namely adivasi, has been in wide use among politicians, social workers, administrators, and even social scientists. Indeed, the term adivasi or indigenous peoples has become an important marker of identity among tribal people in India today.

After India attained independence, there was a major shift in the values and ideology of the country. It moved away from the ideology of inequality towards equality, at least in the public domain. The conferment of citizenship rights on all irrespective of caste, class, religion, or race has been the culminating point of the of equality. Chapter Four examines the impact of this on tribes in terms of the conferment of civil, political, and social rights, and sees in this a means of incorporating tribes into the state structure. It points to the unequal terms of trade between citizenship rights conferred by the state and the economic rights held traditionally by tribes. In return for citizenship rights, the state takes away the most precious rights hitherto enjoyed by tribes, namely the rights over land and forest.

The Indian Constitution has provided safeguards for protecting and promoting the interest and welfare of tribal people, thus also providing space for their empowerment. These constitutional provisions include reservations and special administration in the form of the Fifth and Sixth Schedule areas. In addition, special legislative measures were also adopted to prevent alienation of land from tribes to non-tribes. However, the constitutional provisions have certain limitations, especially in respect of the needs and aspirations of tribal people. Frustration with these limitations led to the demand for greater autonomy, either in the form of a separate state or in the more effective control of resources. The formation of separate states or regional/district councils, or even village councils under the Panchayats (Extension to

Scheduled Areas) Act (PESA), has been explored against this backdrop. These issues have been examined at length in Chapter Five.

The tribals' response to the constitutional provisions and to other state agenda-making practices has been examined in Chapters Six to Nine. One of the ways in which tribals have responded to state policies and programmes has been through various movements. Chapter Six examines the nature of some recent movements in tribal societies and looks critically at some broad generalizations made about them. Most of these movements have been concerned with such issues as culture, language, identity, land, forest, social mobility, and autonomy. Of these movements, the autonomy movement has been subjected to a detailed inquiry since it has been one of the most widespread and well-articulated struggles of tribal people in India. These movements are examined in a comparative perspective not only in relation to movements in north-eastern India but also in relation to the rest of the country.

Chapter Seven deals with the issue of tribal identity in the context of culture, particularly language and religion. It discusses the state policy and state administrative practice towards tribes with respect to their languages and religions, and shows how these (the constitutional provision of protection notwithstanding) have affected the distinct identity of tribal people. The response of tribal people to state policy and state administrative practice has been explored in terms of the increasing demand for recognition of their distinct languages and religions. Before examining this area, the chapter outlines the broad social science perspectives on tribal society in India.

The other issue emanating from the constitutional provision of protection is the question of reservations. Chapter Eight takes this up for critical discussion. It explores the extent to which tribes have been able to make effective use of this provision in the domains of politics, education, and employment, and compares the situation of tribes with that of the scheduled castes. The chapter also examines why the scheduled tribes have not fared as well as the scheduled castes in availing the provision of reservation and provides some tentative explanations.

In view of the damage caused to the ecology and the environment by modern society, there has been a renewed interest in ecological and environmental concerns, in turn prompting a renewed interest in the study of tribal society, since tribes are seen as living in close harmony with the natural environment, and hence as natural conservers of the ecology and the environment. It is against this backdrop that Chapter Nine examines the relation between tribes and forests. It examines this at two levels—existential and cultural—and points to an orientation of rational adaptation rather than rational mastery in the tribal people's relation to the natural world. The chapter then looks at the political economy of forests and discusses the forest policy pursued during the colonial and postcolonial periods in India and the effects this has had on tribes and on their means of livelihood as well as on the ecology and the environment. The chapter also deals with more recent concerns of environmental protection and forest conservation and the effects of these on tribes, especially in relation to their survival and livelihood.

Tribal society has become considerably differentiated and certain forms of inequality have emerged. Of the forms of inequality in tribal societies, the most striking is gender inequality. Chapter Ten looks at this issue, beginning with an examination of the position of women in traditional social settings and then discusses their situation in the changing context of social formations and social differentiation witnessed in tribal society, especially in post-independence India. The issue of gender inequality is a subject of heated debate in many tribal societies today, especially those that have become considerably differentiated. The chapter analyzes the critical issues involved in the debate.

References

Bara, Joseph. 2002. 'Tribe—Beast or Man: Case for a New Concept of Tribe.' *Social Action* 52 (2): 121–33.

Béteille, André. 1986. 'The Concept of Tribe with Special Reference to India.' *European Journal of Sociology* 27 (2): 297–318.

———. 1993. 'Myth of Indigenous People.' *Times of India*, 5 January.

Desai, A. R. 1960. 'Tribes in Transition.' *Seminar* 14: 19–24.

Dumont, Louis. 1998. *Homo Hierarchicus*. New Delhi: Oxford University Press.

Elwin, Verrier. 1944. *The Aboriginals*. Bombay: Oxford University Press.

———. 1960a. *A Philosophy for NEFA*. The Advisor to the Governor of Assam, Shillong.

———. 1960b. 'Issues in Tribal Policy Making.' *Seminar* 14: 25–28.

Ghurye, G. S. 1963. *The Scheduled Tribes*. Bombay: Popular Prakashan.

Government of India. 2001. *Report of the Working Group for Empowering the Scheduled Tribes During the Tenth Five Year Plan (2002–2007)*. New Delhi: Ministry of Tribal Affairs.

Government of India. 2001. *Report of the Steering Committee on Empowering the Scheduled Tribes*. New Delhi: Planning Commission.

IIPS (International Institute of Population Studies). 2000. *National Family Health Survey.*

Khubchandani, L. M. 1992. *Tribal Identity: A Language and Communication Perspective*. Shimla: IIAS.

Pathy, J. 1988. *Ethnic Minorities in the Process of Development*. Jaipur: Rawat Publishers.

Ray, Niharranjan. 1972. 'Introductory Address', in K. S. Singh (ed.), *Tribal Situation in India*. Shimla: IIAS, pp. 3–24.

Risley, H. H. 1903. *Census of India Report*. Simla: Government of India Press.

Roy Burman, B. K. 1994. *Tribes in Perspective*. Delhi: Mittal Publications.

Sengupta, N. (ed.). 1982. *Fourth World Dynamics: Jharkhand*. Delhi: Authors Guild Publications.

Singh, K. S. 1993. 'The Problem in Marginalized Tribals.' *Seminar* 412: 1–7.

Vidyarthi, L. P. 1982. 'Research on Tribal Culture of India', in P. K. B. Nayar (ed.), *Sociology in India: Retrospect and Prospect*, edited by P. K. B. Nayar. Delhi: B. R. Publishing Corporation, pp. 351–438.

Chapter 2

The Transformation of Tribes: The Terms of Discourse

There are more than 400 groups in Indian society officially designated as scheduled tribes. All these groups have been undergoing changes, observed and described by various people for nearly a century. The consequences and implications of these changes, however, have been seriously misconstrued. The conventional wisdom among anthropologists has been that when a tribe undergoes change resulting from the end of its isolation and its integration with the larger society, sooner or later, and with unfailing regularity, it becomes a caste. While this may have been true to a greater or lesser extent until the 1930s and 1940s, the argument is no longer valid. Yet anthropologists have continued making such a generalization. Moreover, this generalization is based on an inadequacy of data, concepts, and arguments. While tribes continue to undergo changes—and these changes are of many kinds—these no longer transform the tribes into castes. For example, the Oraons today practice various religions and speak more than one language; they earn their livelihoods from a variety of occupations, both agricultural and non-agricultural. Yet, despite this enormous differentiation, they are still Oraons in some socially significant sense. They have not become a caste with any definite standing in the social hierarchy of the caste system. This argument has implications not only for an understanding of tribes but also for an understanding of Indian society as a whole. The most important implication is that new castes are no longer being formed, whether by the transformation of tribes into castes or by any other means. Tribes have, of course, become peasants and they have also been socially differentiated, but, contrary to conventional wisdom, this process has occurred without any loss of their own distinct identities.

Caste and Tribe

[The hallmark of Indian society has been described as diversity or heterogeneity. Of the distinguishing characteristics of Indian society, religion, language, region, caste, and tribe have been considered the most important. However, none of these categories has been as contentious, conceptually and theoretically, as the category of tribe.] It has generally been said that the categories of religion, language, region, and caste in India have been rooted in the consciousness and the social relations of the people at large. They have also existed in history over large swathes of time. Such has not been the case with the category of tribe. The latter was added to the list mentioned above by the British in the nineteenth century. The category of tribe is hence seen as a colonial construction (Béteille 1995; Singh 1993). In an attempt to understand Indian social reality, the category of tribe has nevertheless come to be used extensively in social science literature in general, and in sociological and anthropological literature in particular.

At the time when the British began to write on Indian society, the term tribe was used in more than one sense in general parlance. In one sense, it was used to refer to a group of people claiming descent from a common ancestor. In another sense, it was used to refer to a group of people living in primitive or barbarous conditions; this latter sense emerged during the colonial period. The usage in the former sense has a longer history than the usage in the latter sense, when the term came to be associated with the colonial encounter. Yet the sense of tribe that developed later (namely the primitive or barbarous living conditions) has come to dominate anthropological writings. The concept of tribe has thus undergone changes in the course of history. Early British writings on India did not study groups or communities from the perspective of the caste/tribe category, which was a later development. The Asiatic Society of Bengal, which is credited with piloting the studies of Indian society, defined the scope of its mission as the study of 'nature' and 'man'. Hence, the sense in which British ethnographers used the term tribe in India, especially in the early phase, is not very clear. The impression one gets is that the usage of the term in the sense of common ancestry may have been more common than the usage in the sense of primitive or barbarous living conditions. The usage of terms such as Rajput, Ahir, or Jat tribes, as well as the synonymous use of the terms tribe and caste in eighteenth-century writings on India, tend to support such a view. The two terms were also used in a cognate manner, as one can see in Risley's use of the phrase 'tribes and castes of India'. This shows that ethnographers had difficulty differentiating one from the other, at least in the initial stage. Perhaps there was little empirical evidence that could be of help in this regard. Anthropologists may also have lacked conceptual clarity in their minds as to the categories of caste and tribe.

That the latter was also the case is obvious from the way in which administrative-cum-census officials handled the two categories in the census reports of 1881, when the first 'proper' all-India census was undertaken. The term used there was not 'tribe' but 'forest tribe', and that too as a subheading within the broader category of agricultural and pastoral castes. A somewhat more serious effort at drawing a distinction is reflected in the later censuses. Risley and Gait, who were in charge of the 1901 and 1911 censuses respectively, added the category of 'so-called animists' to the list of caste and others, and included against each the number of people professing to follow Hinduism and animism. Marten followed the same pattern in the 1921 census, except that he changed the heading from 'animism' to 'tribal religion'. Hutton continued with the distinction between tribes and others in terms of religion. Tribes were distinguished from others by the type of religion they practiced, such as Hinduism, Christianity, Islam, etc. It is interesting to note that tribes were distinguished from others not on the basis of caste, or caste-like features, but religion. For Hutton, then, the tribe–caste distinction can be maintained only on the basis of religion. Tribes were thus defined as those that practiced 'animism'. Of course, officials in charge of the census operation were not satisfied with this basis for the demarcation of tribes. They were of the view that it was difficult to distinguish the religion of the tribes from the religion of the lower strata of Hindu society. Keeping in mind these observations, Ghurye (1963: 205) went to the extent of describing the tribes as 'Backward Hindus'. He writes that so-called aboriginals who form the bulk of the scheduled tribes, and who have been designated in the censuses as Animists, are best described as Backward Hindus. The criteria worked out for making such a distinction were thus far from satisfactory.

In the post-independence period, there was a greater concern about, and more systematic efforts aimed at, distinguishing tribe from caste. Yet even today, scholars have not arrived at a systematically worked out criterion. In general, they have tried to distinguish one from the other on the basis of many criteria. It has generally been assumed that the two represent two different forms of social organizations. Caste has been treated as a system regulated by the hereditary division of labour, hierarchy, the principles of purity and pollution, and civic and religious disabilities. On the other hand, tribes have been characterized by the absence of features attributed to the caste system. The two types of social organizations are also seen as being governed by different sets of principles. It is said that kinship bonds govern tribal societies. Hence, all individuals are considered to be equal. Lineage and clan tend to be the chief units of ownership as well as of production and consumption. In contrast, inequality, dependency, and subordination are seen as integral features of caste society. It is also said that tribes do not differentiate between the utilitarian and the non-utilitarian functions of religion as sharply as caste groups do. Caste groups tend to maintain different forms, practices, and behaviour patterns for each of these two functions of religion. In contrast, tribes maintain similar forms, practices, and behaviour patterns for both these functions of religion. Tribes and castes are also shown to be different in respect of the psychological disposition of their members. Tribes are said to take a direct, unalloyed satisfaction in the pleasures of the senses, whether in food, drink, sex, dance, or song. As against this, people in caste-based societies are seen as having ambivalent attitudes towards such pleasures. Further, in jati society, the village is expected to be culturally heterogeneous, with each jati following a unique combination of customary practices. Tribes, on the other hand, expect their society to be homogeneous, or at least not necessarily heterogeneous (Mandelbaum 1970: 577). Based on these descriptions, certain images and propositions have been put forward with respect to the concept of tribe in India. These include the absence of exploitative classes and organized state structures; multifunctionality of kinship bonds; all-pervasive religion; segmentary character of socio-economic units; frequent cooperation in the pursuit of common goals; shallow history; distinct taboos, customs, and moral codes; youth dormitory; low level of technology; common name, territory, descent, language, culture, etc. (Pathy 1992: 50).

Paradoxically, however, these sets of denominators on the basis of which tribes are differentiated from non-tribes (that is, castes) are not subscribed to by a large number of groups identified as tribes in India. Even groups that do subscribe to these attributes do not stand in the same or similar relation to each other with respect to these attributes. At one end, are groups that subscribe to these features in toto; at the other, are those groups that hardly exhibit these attributes. The large majority, however, stand somewhere in between the groups subscribing to the attributes referred to above to a greater or lesser degree. Hence, assumptions relating to tribes have frequently proved to be misleading and fallacious. Disregarding the differences among groups in relation to the subscription to such attributes, they have all been identified as tribes. The only thing that they seem to share in common is, as Béteille puts it, that they all stand more or less outside of Hindu civilization. Since the identification of tribes has political and administrative implications, little effort has been made to examine it critically. Rather, the category of tribes generally has been accepted uncritically by social scientists.

Transformation of Tribes into Castes

In colonial ethnography, the concern shown by British administrator–scholars about marking off tribe from caste also led to the rise of a particular conception of tribe. That is, tribes were groups that lived in isolation, and hence had no interaction with the larger population. In contrast, the main concern in postcolonial ethnography has been to show the close interaction of tribes with the larger society or civilization. The relation has, of course, been conceptualized in different ways. Sinha (1958) views tribes as a dimension of the Little Tradition, which cannot be understood adequately unless it is seen in relation to the Great Tradition. In contrast, Béteille (1986: 316) sees tribes as remaining outside of the state and civilization in those contexts where tribe and civilization coexist, as in India and the Islamic world. Thus, though the distinction is maintained, the two are treated not as being isolated but in interaction with each other. Even when tribes have been conceived of as remaining outside the state, which has been the case most often, they have not been treated as falling outside the influence of civilization. Hence, tribes have been viewed as being in constant interaction with civilization. Consequently, tribal society has not been seen as static, but as engaged processes of change.

One of the dominant ways in which the transformation of tribal society has been conceived is in terms of tribes moving in the direction of becoming a part of civilization by being absorbed into the society that represents civilization. Historians and anthropologists have made such an observation in the context of the past. Kosambi (1975) has referred to tribal elements being fused into the general society. Similarly, Bose (1941) has referred to tribes being absorbed into Hindu society. Such a claim has not gone unchallenged. Many anthropological works of the post-independence period still point to phenomena such as tribes being absorbed or assimilated into Hindu society or tribes becoming castes. Tribes are said to have accepted the ethos of the caste structure and to have become absorbed within it. They are treated as being hardly differentiable from the neighbouring Hindu peasantry. Some well-known tribes in this category are said to be the Bhils, the Bhumijs, the Majhis, the Khasas, and the Raj-Gonds. In fact, much of the social anthropological discourse on tribes has been couched primarily in terms of tribes being transformed into castes. Nowhere is this better reflected than in the classification of tribes provided by eminent anthropologists. The classification has been made differently by different scholars, but all invariably refer to a stage of incorporation into Hindu society. Roy Burman (1972) in his earlier work classified tribes as (1) those incorporated into Hindu society; (2) those positively oriented towards Hindu society; (3) those negatively oriented towards Hindu society; and (4) those indifferent to Hindu society. Vidyarthi and Rai (1977) talked of tribes as (1) those living in forests; (2) those living in rural areas; (3) those who were semi-acculturated; (4) those who were acculturated; and (5) those who were assimilated. The criteria for classification used by Vidyarthi and Rai suffer from a lack of logical consistency since some categories in the classification are based on ecological consideration and others on level of acculturation. Elwin (1944) divided tribes into four categories. These were (1) the purest of the pure tribal groups; (2) those in contact with the plains and therefore undergoing changes but still retaining the tribal mode of living; (3) those constituting the lower rung of Hindu society; (4) those adopting the Hindu faith completely and living in a modern style. Elwin even went to the extent of arguing that the whole aboriginal problem was one of how to enable the tribesmen of the first and the second classes to advance directly into the fourth class without having to suffer the despair and degradation of the third. Dube (1977) also classifies tribes along almost the same lines as those described by Elwin. Many others, including Bose (1941, 1971) and Fuchs (1973) have not made specific classifications *of tribal population* but do state that tribes occupy either the lower or the higher rungs of the caste hierarchy by being absorbed into Hindu society.

Some scholars caution us about such a conception of tribal transformation. Roy Burman (1983, 1994) in his later writings argues that if the transformation of tribe into peasant is not to be taken for granted, then the transformation of tribe into caste in the Indian context can also hardly be taken for granted. This he does by providing a critique of the Bose and Srinivas models. He points to a lack of protection from the system of caste for Hinduized tribes, the rationale for the transformation of tribe into caste in Bose's model, and the empirical reality of contra-Sanskritic movements against Srinivas's model of Sanskritization. Pathy (1992: 50–51) questions the dominant trend of understanding the transformation of tribe into caste on account of a lack of historical and contextual evidence. Yet he endorses Kosambi's observation that the entire course of Indian history shows the fusion of tribal elements into the general society.

The transformation of tribes into castes is seen to occur through certain methods, which have been conceptualized in different ways. Kosambi (1975) considers the adoption of the technology of Hindu society by tribes as the main method of integration into Hindu society. Bose (1941) talks of the Hindu method of absorption that takes place under the mode of production. He says that tribes are drawn into the system because they find protection within it, the system being non-competitive. Sanskritization is also seen as a method of absorption. The other significant method of tribal absorption into Hindu society is what Sinha (1962, 1987) calls state formation. He states that the processes of acculturation, Hinduization, and social stratification in village society cannot be understood properly unless the data are examined in the broader context of the formation of principalities. He argues that state formation provided the definitive socio-political framework for the transformation of the tribal system into the regional caste system.

Sanskritization

As noted earlier, it has generally been held that tribes in contact with non-tribes have been undergoing change, and that this change has been in the direction of absorption into Hindu society through complex social processes. Scholars have conceptualized the processes at work in various ways. This is evident from the range of terms used to describe these processes, the most common terms being Sanskritization and Hinduization. Anthropologists have also used such specific terms as 'Kshatriyaization' and 'Rajputization' in place of Sanskritization. These terms describe different social processes at work, though in reality these processes coincide and overlap. Despite this, there has been a tendency among social scientists to use them interchangeably or synonymously. Often the difficulties arising from the use of such terms are overcome by the use of such generic terms as acculturation, assimilation, and absorption. However, the main processes through which the transformation of tribe into caste is understood are Hinduization and Sanskritization.

Do processes such as Hinduization and Sanskritization that occur in tribal society lead to its dislocation and pave the way for its absorption into Hindu society? Does tribe by virtue of becoming acculturated cease to be tribe and become caste? Almost all the scholars referred to earlier tend to agree that this is the case. To these scholars, tribes eventually cease to exist as entities independent of the caste society from which they were earlier differentiated. While this may have been the case in the past, it is no longer true of India in the post-independence period.

Since the acculturation of tribes to Hindu society, or the transformation of tribes into castes, is attributed to the processes of Sanskritization and Hinduization, it is important at the outset to examine the appropriateness of these terms and concepts, especially Sanskritization. Sanskritization is seen as a process whereby the communities placed lower on the caste hierarchy emulate the way of life of the dominant caste of the region. Through this process of emulation, they are able to move up the caste hierarchy. At times, more specific terms such as Kshatriyaization and Rajputization have been used to describe this process. This process was used to understand and describe the dynamics of social change in caste society. Sociologists and social anthropologists have extended this term and concept to also describe a certain process of change taking place in tribal society. Is it valid to extend the meaning of the term and concept to describe the transformations being witnessed in tribal society? In my view, the extension, in the sense in which it is being used, is far from appropriate in the context of tribal society. It is inappropriate because it assumes that tribes are, first of all, part of Hindu society, and, second, that they are part of caste society. Tribes have, however, been conceived of as tribes precisely because they are outside Hindu society as well as outside caste society. Can there be a process of Sanskritization as it has been conceived without a tribe first becoming part of caste Hindu society? The question that arises is whether Hinduization is the same as Sanskritization. The two are, of course, interrelated, but it may be more appropriate to describe the process involved in the context of tribes as Hinduization rather than Sanskritization. This is so because climbing up the caste hierarchy is not the overriding concern among tribes. Of course, it is not possible to conceive of the Hindu faith and practices outside the caste structure. Hinduization thus invariably entails the assumption of a caste status. But the caste status that is accorded to tribes is said to be a low-caste status. If this is the case, where is the process of social mobility in the case of tribes? What is it that the tribes gain through this process? They have not made a claim to or a demand for a higher caste status (Hardiman 1987: 158–59).

Rather, it is outsiders who impose such a status on tribes. In fact, even after undergoing Hinduization, tribes by and large remain outside the hierarchical structure of Hindu society. If at all there have been such claims, these have been made, as we shall see, only after the tribes have been drawn into the larger social structure of the neighbouring Hindu and linguistic community. The case of the Meteis and the Koch-Rajbongshis may be taken as points of reference. Unlike others, the two communities had taken to Hinduism as a whole. Yet it was not clear what caste status and caste name they assumed after the adoption of Hinduism. They claimed Kshatriya status, but this claim was made long after their adoption of the Hindu way of life. Moreover, the claim was made for the entire community, and not for a segment of it. Elaborate caste differentiation did not exist among them. If at all there were Brahmins, they were immigrants. In Manipur, they were not from among the Meteis, but belonged to other ethnic communities and were not considered part of Metei society. The latter too saw themselves as different from the Meteis. This has also been the case with the Koch-Rajbongshis, who had not only embraced Hinduism but had also adopted the Bengali and Assamese languages. Yet their integration into the dominant regional community had been far from complete. In fact, they were addressed and identified more by their ethnic names than by their caste name. Indeed, it is not certain that they have any caste name at all. That they have been making claims to Kshatriya status is an entirely different story.

The problem with the concept of Sanskritization in the case of tribes does not end here. There is also a problem of the reference group. It is far from clear from the literature as to which caste groups the tribes (barring those with a royal lineage) emulated in their respective regions. The royal lineage has invariably emulated the Rajputs in the Hindu caste structure and has entered into matrimonial alliances with them. Thus, whereas the upper strata of tribal society became integrated into Hindu caste society, the subjects under their kingdom continued to live outside Hindu society, though the process of Hinduization may have taken place among them. Climbing the ladder of caste hierarchy was not their main concern. Given this, it would perhaps be more appropriate to speak of Hinduization rather than Sanskritization in the context of tribes in India. Further, if at all tribes consider some castes as superior, this is not because of their caste status but because they happen to be jagirdars, thicadars, lambadars, etc. The question that may be asked then is why tribes Hinduize themselves even though they attain no higher status. Do they want to be absorbed into the larger society? This may have been the case in the past, but it is no longer the case today. Today, the process of acculturation to the ideas, values, and practices of the dominant community is more a matter of being like the dominant community than a matter of being part of that society by assuming some kind of caste status within it.

Hinduization

It is thus apparent that tribes have been described as castes more in reference to Hinduization than Sanskritization. Indeed, this seems to be more often than not the basis for describing tribes as castes in the social anthropological literature. And yet can they be described as castes just because they have become Hinduized? Is the process of Hinduization sufficient for designating a group as a caste? Is it not possible for a tribe to be Hinduized and yet remain outside the caste system, namely to be governed by the social organizational principles of a tribe rather than those of a caste? Questions such as these have either not been given sufficient attention or have been overlooked in studies in which tribes have been modelled on the caste system or incorporated into the civilizational framework. If Hindu society cannot be understood outside caste society, then the transformation of tribe into caste or Hindu society, as scholars have been arguing, is problematic. Indeed, the whole argument for the transformation of tribe into caste seems to be misplaced, and even erroneous.

Theoretically, it is possible to become Hindu in the sense of embracing a form of the Hindu faith and practices without becoming part of Hindu society in the sense of accepting Hindu social organization, namely caste. If, on the other hand, Hindu society and caste organization are inseparable, then Hinduization alone cannot account for the transformation of tribe into caste. In fact, there are other aspects to which sociologists and social anthropologists need to pay attention. Do tribes actually become a part of the structure of caste society after they have taken to the process of Hinduization/Sanskritization? What caste name do they bear? What position do they occupy in the caste hierarchy? It is also not clear whether the groups involved in the process occupy the same position, or whether there is a hierarchical arrangement among them as has been the case with the untouchables. Also, what caste roles do they assume, say, in villages where both tribes and castes live, as in most villages of Chhotanagpur, where caste groups like the Banias, the Brahmans, and the Rajputs live in the same village as the tribals? In fact, the nature of the interaction between tribes and castes is governed more by market forces and the force of economic interdependence than the purity–pollution principles. Further, their life in the village continues to be grounded on the principle of kinship and on the absence of hierarchical ordering. In short, tribes do not enter into any kind of social, cultural, and ritual dependence within the caste structure of Hindu society even after their acculturation to the Hindu belief system and practices. It is doubtful, then, if it is appropriate to study people described as tribes from the perspective of the caste structure. This is precisely, however, what anthropologists have done. They have tried to find caste where it does not exist.

It should also be noted that tribes have not moved into processes like Hinduization or Sanskritization as entire groups. The general pattern among them is that only a section adopts a new pattern of life provided either by Christianity or Hinduism or Islam. If this is the case more often than not, can we describe some people of the same group as a caste and others as a tribe? Can one and the same group be caste and tribe at the same time? The social situation of a village where tribes form a minority and where they are absorbed into Hindu society is extended to villages and regions where they may not be a minority and where, even if the process of Hinduization is taking place, they may not abandon their old affinities and identities. However, where tribes have taken to Hinduization as a whole, they have, to a great extent, moulded themselves along caste lines. They have even identified themselves as castes, and others too have addressed them as castes rather than as tribes. The Koch-Rajbongshis of Assam and West Bengal, referred to above, may be taken as a case in point. But the phenomenon of the group as a whole adapting to a different value system is rather rare. Even where such a thing has happened, it has not given rise to a hierarchical caste structure. The group as a whole tends in general to belong to the same stratum. Nor is the group adequately integrated into the caste structure of the neighbouring regional community.

In examining the question of the transformation of tribe into caste, it is not enough to limit the discussion only to the relationship between tribes and caste society. There is also a need to look into the question of how tribes themselves perceive, identify with, and relate to caste society. How did tribes perceive themselves after they had taken to certain aspects of Hindu beliefs and practices? Did they perceive and identify themselves as tribes or as castes? The important ways through which tribes took recourse to the process of Hinduization or Sanskritization are what anthropologists have described as a 'religious/cultural movement'. This movement has been described more popularly as the Bhagat movement among tribes. In fact, despite the process, of Hinduization or Sanskritization, tribes do not set aside a section of themselves as caste and another as tribe. The Hinduized tribes are not treated as having moved away from the status of tribe. Rather, tribes are categorized into different groups depending on the type of religious values they have chosen to adopt. They are, therefore, variously described as Christians, Bhagats, Sarnas, etc. It is interesting to note that tribes, even when they have been Hinduized, do not describe themselves as Hindus but as Bhagats. It is outsiders, census officials, and anthropologists who have tended to describe them as 'Hindus'. Anthropologists have even been prone to describing them as castes. Tribes, however, do not identify and designate themselves as belonging to different castes in the sense used and understood by outsiders and social scientists. Nowhere is this aspect of distinctive identity more evident than in the movements launched by tribes, especially those pertaining to autonomy, land, forest, and employment. In these movements, the divide between caste and tribe has been relatively sharp. And yet tribes that have been Hinduized have shown solidarity with groups described as tribes as against those belonging to the caste category. In short, the process of Hinduization is necessary but not sufficient for tribes to be integrated into the structure of Hindu society, namely caste society. To be integrated, tribes must be drawn into the social organization of the caste. However, this is generally not an empirical reality.

Language

The foregoing discussion points to the fact that it is not possible to become a caste without first being integrated into the structure of Hindu society. Where such integration does occur, a very important process is the acculturation of tribes to the language of the regional community.

It is interesting to note that castes as a form of social organization are operative only within a linguistic community. Hence, it is possible for tribe to become caste only after it has been assimilated into the regional linguistic community, such as the Bengali or the Oriya or the Assamese communities. This process of acculturation—which is central to the integration of tribe into the regional community, and therefore into caste society—has unfortunately been glossed over by sociologists and social anthropologists. In fact, it is not possible to become integrated into caste society without first becoming integrated into the linguistic community. Correspondingly, tribes were not only differentiated in opposition to castes but also in opposition to the dominant community of the region. The dominant community was invariably a linguistic community. Besides representing a different language, it also represented a different religion, customs, social organization, and way of life. This raises an interesting question, namely whether tribes after they have become Hinduized, and even 'caste-like', should be treated either as castes or as tribes if they continue to maintain their language. After all, as noted earlier, tribes have also been seen in opposition to language or the dominant linguistic community. Can they be both tribes and castes at the same time? This seems far from tenable, as the two not only constitute different linguistic communities but also contrasting types of social organizations. Are tribes, then, to be treated as castes just because there has been a process of Hinduization at work among them? Do tribes by Hinduizing and Sanskritizing themselves become castes while retaining their language, culture, customs, social practices, and so on?

What is important to note here is the influence of Hinduism or Hindus on tribes. This, however, does not make them Hindus. To be Hindus, they need to be drawn into the structure of Hindu society, which is possible only by being drawn into the structure of the regional linguistic community. Tribes are differentiated from non-tribes not on the basis of religion alone. That is what colonial ethnographers argued. Anthropologists have differentiated tribes from others, however, on the basis of more than one criterion. Of these criteria, the most important have been language and social organization. Tribes have thus been treated as tribes precisely because they have been outside the dominant regional community, and thereby outside the complex of civilization. However, even with changes at the level of culture, including religion and language, one is still not sure if tribes could be said to have become castes. Of course, much was dependent on the nature of their linkage with the social structure of the regional linguistic and Hindu society. Indeed, what seems to me to be the most crucial prerequisites for the integration of tribes into the structure of the regional community are not only religion and language but also the organizational structure of the regional community. Of all the aspects of acculturation, acculturation to language is, to my mind, more important than acculturation to religion, though the latter is not altogether unimportant. Unfortunately, sociologists and anthropologists have never given language the place it deserves in understanding the process of the transformation of tribe into caste. And yet without going into these questions, anthropologists have jumped to the conclusion that tribes are becoming castes or are becoming integrated into Hindu society. It is also to be noted that being drawn into a larger society does not mean that tribes cease to exist as a society. Do they, then, by virtue of cultural change occurring within society cease to be a society? Does Bengali society cease to be a society in the wake of the processes of westernization and modernization occurring within it? The paradox is that nobody ever denies the existence and identity of Bengali society, but if the same process occurs within tribal society, the general trend is to negate its existence. The reality that tribes exist as a society as much as other societies exist is, unfortunately, denied to tribal society when it undergoes change in the context of its encounter with the larger society. Anthropologists have been quick to incorporate them into the larger society at the slightest sign of change among them. In their zeal to emphasize change or the acculturation process, the aspect of continuity—about which anthropologists talk so fondly in the context of the larger Indian reality—has been completely overlooked in the context of tribal social reality.

This discussion shows that the conclusions reached by sociologists and social anthropologists are based on an inadequacy of ethnographic data, concept, and even logic. There is hardly any inquiry into the ways in which the Hinduized tribes are linked with castes and the ways in which they relate with in their own group. In addition, no effort whatsoever has been made to ascertain if the acculturated tribes were regulated by the principle of caste organization or by the principle of tribal organization. Issues such as these, central to the argument in support of the transformation of tribe into caste have, unfortunately, not been adequately examined. In addition, even concepts such as Sanskritization and Hinduization used for understanding the transformation of tribe into caste are inadequate for advancing an argument in support of such a claim.

Tribe and Peasants

Tribes in India have not only been studied in relation to caste but also in relation to peasants and peasant society. Peasants and peasant society in social anthropological literature have, however, invariably been conceptualized as being in contrast to tribal society. A distinction is thus maintained not only between tribe and peasant but the two are also seen as different types of societies. Tribes in anthropological literature generally have been defined as communities that are more or less homogeneous, having a common government, a common dialect, and a common culture. It is, however, as Béteille (1960) puts it, one thing to show boundaries between tribes and non-tribes, or between different tribes, and quite another thing to specify the characteristics of tribal societies as such. An attempt has, therefore, been made to pay attention to the latter. Accordingly, tribes have come to be defined by the features of a segmentary system. This means that tribes are not only conceived as being small in scale but also as being representative of a definite structural type that is quite different from the more complex social system in which the peasantry and the gentry coexist. Ideally, then, tribal societies are small in scale, are restricted in the spatial and temporal ranges of their social, legal, and political relations, and possess a morality, religion, and world view of a corresponding order. In short, tribal societies are self-contained units. In contrast, peasants and peasant society are not seen as a whole society, but as a part society with a part culture. Redfield (1956), for example, uses the term peasant for any society of small producers who produce for their own consumption through the cultivation of land and who form a segment of a town-centred economy and society. Similarly, Shanin (1973) defines peasants as small agricultural producers who, with the help of simple equipment and the labour of their families, produce mainly for their own consumption and for the fulfilment of obligations to the holders of political and economic power. Foster (1953) too views folk culture and folk society not as a whole society, or as an isolate, but as a 'half society', a part of the larger unit. The 'half society', he says, has a symbiotic spatial–temporal relationship with the larger cultures of which it is a part. He views folk culture as a common way of life that characterizes some or all the people of many villages, towns, and cities within a given area. Folk and urban are thus not polar opposites, but rather both form integral parts of certain types of socio-cultural units. This being the case, reference is made to three types of societies in the Indian context, namely tribal, peasant, and urban.

There has been much inquiry in anthropology into the extent to which tribal people in India can be regarded as peasants. The inquiry arises from the fact that of the peoples or communities described as tribes, not all stand at the same level of development. Accordingly, they have been classified into various categories depending on their characteristic mode of livelihood. Bose (1971: 4–5), for example, divides tribal people into five categories. These are: (1) hunters, fishers, and gatherers; (2) shifting cultivators; (3) settled agriculturists using the plough and plough cattle; (4) nomadic cattle keepers, artisans, and agricultural labourers; and (5) plantation and industrial workers. Of these, some are considered to be no different from the non-tribal population, which has been described mainly as a peasant population. The process of peasantization among the tribes in Indian history is attributed largely to cultural contact with the non-tribal world. The process has also been attributed to the development strategy of the Indian state, especially in the post-independence period. Tribal societies are, therefore, no longer seen as static but as being in the process of change.

The case for the transformation of tribes into peasants has, of course, been made keeping in mind different considerations by different scholars. Some have done so keeping in mind the fact that tribes have moved away from hunting and/or fishing or shifting agriculture to terraced or settled agriculture, and some have done so on the ground that tribes have moved to plough agriculture. In fact, more often than not, tribes have been described as peasants without even adequately applying the criteria used for defining peasants. Others describe them as peasants because they see little difference between the way in which tribes make their living and the way in which the larger non-tribal community makes its living. There are, of course, scholars who have tried to look into the problem by systematically applying the criteria evolved in anthropological writings. Thus, keeping in mind the segmentary system in terms of which tribes have been defined generally, Bailey (1961) differentiates tribes from castes or peasants. It is worth noting here that Bailey was more interested in differentiating tribe from caste rather than in differentiating tribe from peasant. He characterizes caste society as predominantly hierarchical and organic, and tribal society as basically segmental. Sinha (1965) points to the inadequacy of such a characterization. He says that there are some parts of India where peasants, particularly those belonging to the Rajput and the Jat castes, approximate more closely to the characteristics of tribes. He even goes to the extent, as noted earlier, of viewing tribes as a special case of the Little Tradition in the civilization of India. Béteille (1974: 61) relates the concept of peasant as formalized in Shanin's definition to the empirical realities of the tribes in Chhotanagpur, and shows that the realities there more closely approximate the concept of peasant than the realities obtaining elsewhere among communities that generally have been described as peasants.

With a shift in focus from tribe to peasant, there has also been a shift in the study of tribes. The study of tribes as communities has given way to 'village studies'. Indeed, village studies are seen as being different from, or as an alternative to, tribal studies. There is little doubt that this way of contrasting tribal studies with village studies is a direct consequence of the false opposition posed between tribe and peasant in anthropological writings. It is important to note that the dichotomy posed between caste and tribe in the study of Indian society has also led to a dichotomy between tribe and peasant. It is interesting to note that Indian society has not only been seen as a caste society but also as a peasant society. The two, in fact, have been seen as coterminus. Conversely, communities identified as tribes are not treated as peasant societies and are assumed to make a living in ways that are different from those of the larger caste society. Correspondingly, tribes in India are not only contrasted with the caste dimension of Indian society but also with the peasant dimension. Hence, any tribal community that has been making a living in the same way as that of the larger community is said to be either in the process of becoming or as having already become a peasant society. This means that the tribal community is either in the process of ceasing to be a tribal society or it has already ceased to be one. Does this mean that there is nothing left of the attributes associated with a tribe in the changed situation of the peasantization process? Now one is confronted with such a problem because of the dichotomy posed between tribe and peasant. The dichotomy is indeed false. Tribe can still move in the direction of the peasantry without losing the attributes of tribe as a society. Social scientists have, therefore, not been quite at ease when they talk of the transformation of tribe into peasantry. Oommen (1995: 21–37), for example, points out that with the advent of settled agriculture among tribes, they have become increasingly specialized as peasants, but even the settled agriculturists among them are not yet peasants in several respects, and certainly do not share the folk version of the elite culture. It has further been stated that while there is a common economic and political denominator, the culture is distinct. The culture of the rural Hindu is the folk version of the elite culture; they are not the same, though they are interrelated. There is no such mutual dependence between tribes, on the one hand, and the folk and the urban, on the other hand. Thus, the problem encountered with caste also appears in the case of peasants.

Social Differentiation

There is a third framework of reference in terms of which tribes in India have been addressed and studied, and this is social differentiation. Sometimes this has also been couched in terms of class or social stratification. Tribal society has never been static, but change has never been as unprecedented and dramatic as in the last fifty years. As a result, tribal society has moved from homogeneity, the hallmark of tribal society, to considerable heterogeneity. To begin with, there is occupational differentiation in tribal society. One can find in the same society people who are engaged in agriculture (shifting or settled) or in trade and commerce or even business. Others work as landless agricultural labourers, quarry or mine workers, stone crushers, and plantation and industrial workers. Still others make their living by joining a profession, such as lawyers, doctors, teachers, government servants, politicians, etc. Along with occupational differentiation, there are also differences in wealth and income, giving rise to social stratification in the form of class, not only in the quantitative but also in the qualitative sense. There have also been differences of religion, ideology, values, political orientation, way of life, etc. among members of the tribal community. In view of this, it is generally held that a given tribal society has become like any other component of Indian society, and hence that the society is no longer a tribal society and its members no longer tribes.

Basis for Misconstruction

The concepts of caste, peasant, and social differentiation against which tribes have been studied have invariably led scholars, as noted above, to state or conclude that tribes are becoming castes, or peasants, or socially differentiated groups. What this in effect means is that tribes have become like other sections of Indian society, and as such have ceased to be tribes; they have lost the traits and features that made them tribes in the first place. This thinking has led to the conclusion that tribes and tribal societies by becoming castes, or peasants, or socially differentiated groups cease to be tribes or tribal societies. There is, then, something teleological involved in the study of tribes and tribal societies in the Indian context.

We already know the direction in which tribal society is moving. Nowhere is this more obvious than the background against which tribal society has been studied. Elsewhere in the world where tribes are not linked with the civilization at complex, such conceptual problems do not arise as tribes there are studied in their own right and against the backdrop of the processes at work in those societies. They are not studied against the end point represented by the communities that are part of civilization, as has been the case with tribes in India. Hence, whereas elsewhere the focus has been on how tribes are changing and how they are becoming nationalities or nations, the focus in India has been on how tribes are becoming castes, peasants, and socially stratified groups. It is a corollary, then, that tribes cease to be tribes or tribal societies. And since these are the features that supposedly characterize general Indian society, tribes are viewed as having been absorbed into general Indian society. What follows is that tribes, through these processes, cease to be tribes, and thereby cease to constitute and maintain a separate society and identity. Tribes are being studied primarily from the perspective of how they are being drawn into the mainstream of civilization. The end result is that the reality of caste remains intact but that of tribe disappears, and this process will increasingly spread as tribal societies come into intensive contact with the outside world. Such an empirical and conceptual scenario in the study of tribes exists in India precisely because of the way in which (1) tribes have been conceptualized in anthropological literature; and (2) the frame of reference in terms of which they have been investigated. In a nutshell, tribes have not been studied in their own right but only in relation to general Indian society, which was marked by the features of caste, peasant, and social differentiation.

Now in the conceptualization of tribes in anthropology, three distinct but interrelated elements are seen. First, tribes in anthropology are invariably seen as societies. They are societies like all other societies. That is, a tribe is made up of people; it has boundaries (people either belonging or not belonging to it); people belong to society by virtue of certain rules, which they accept and which impose on them regular and determinate ways of acting towards one another. A tribe as a society is defined on the basis of its boundaries, which set certain limits on the legal, political, economic, and social relations of its members. These boundaries have been defined linguistically, culturally, and politically by anthropologists.

Second, tribes are also seen as a type of distinct society. Godelier (1977: 30), for example, sees tribal societies as being characterized by certain positive and negative features. The negative features are marked by the absence of the positive traits of modern society, namely non-literate, uncivilized, non-industrialized, non-specialized, etc. The positive features are those that are absent from modern society, namely social relations based on kinship bonds, all-pervasive religion, frequency of cooperation for the achievement of common goals, etc.

Third, tribes are also seen as representing a particular stage in socio-political evolution. It is assumed that with the passage of time they will move to a new stage, such as nationhood or nationality.

While these three distinct aspects have gone into the making of the concept, the last two have overshadowed the first to which tribes owe their separate and independent existence. In the process, tribes have been seen primarily as a stage and a type of society. They are seen as representing a society that lacks the positive traits of modern society. To put it differently, they are seen as constituting primitive, simple, illiterate, and backward societies. As the specific features that distinguish tribes change under the influence of education, specialization, modern occupations, new technologies, etc., tribal society is no longer considered tribal society. If the transformation is in the direction of caste, then it is described as having become a caste society. If the transformation is in the direction of peasantization, then it is posited as a peasant society. If the general direction of transformation is towards social differentiation, then it is described as a socially differentiated society. The end result is that tribal society is no longer considered a tribal society, and rightly so, if it is thought of in terms of stages and specific features. But if a tribe is also a society, it is similar to any other kind of society. However, even this reality comes to be denied with the denial of tribal society as such on account of the changed situation.

With this denial, the very existence of tribe as an independent and separate living entity is jeopardized. In the process, anthropologists and other social scientists have overlooked the context in which the term tribe came to be used in Indian society. In the Indian context, tribes were identified and described primarily in terms of being outside civilization. There is, then, something clumsy about the use of the concept of 'tribe' to describe Indian social reality. Such problems may not arise when tribes do not coexist with non-tribal societies. Indeed, problems of the type referred to above could be overcome by the use of the term 'indigenous' people, but not without giving rise to problems of a different kind. Such problems are rooted in the conceptual framework used for understanding the process of transformation in tribal society in India. There is, then, something clumsy and basically wrong about the use of the term tribe in the Indian context.

Tribe as Community

In view of this, it is suggested that the terms of reference for the study of tribes in India should be the same terms used by the tribal people to identify themselves and by which they are known and addressed by other people. It is well known that groups and communities classified as tribes do not identify themselves in these terms (except by the educated), but call themselves by the names of their own tribes, such as the Santhals, the Oraons, the Khasis, and the Garos. This was how the groups now identified as tribes were identified and addressed throughout history. Ray (1972: 8–10) points to this in his introductory essay in *Tribal Situation in India*. He writes that we know that there were janas or communities of people like the Savaras, the Kullutas, the Kollas, the Bhillas, the Khasas, the Kinnaras, and many others whom today we know as 'tribes', bearing names that are recognizable as almost the same even today. Yet the term by which they were known to others was not tribe but jana, meaning 'community of people'. Hence the plea being made here is to study tribes in India in reference to the actual communities to which they belong and which they represent, that is, as Santhals, Khasis, Gonds, etc. If tribes are studied as such, we can overcome the kinds of problems with which we are confronted when we use the term tribe. The transformation occurring in tribal society, whether in the direction of caste, peasant, social differentiation, or religion, becomes meaningful without in any way affecting the identity of the group concerned. The transformation becomes meaningful precisely because tribes, besides being a type of society also constitute a society as such This means that the terms of reference or description in tribal studies should not be caste, peasant, or social heterogeneity, but rather groups or communities, such as regional communities, for example, Bengalis, Assamese, and Gujaratis. The counterparts of tribes are not castes or peasants, as has been the case so far in tribal studies, but communities or societies incorporating castes and peasants, for the latter are not the whole society but only an element of the whole. A tribe, on the other hand, is a whole society like any other society, with their own language, territory, culture, customs, and so on. Hence, as societies, tribes must be compared with other societies and not with castes, as has been the case in sociological and anthropological writings.

References

Bailey, F. G. 1961. '"Tribe" and "Caste" in India.' *Contributions to Indian Sociology* 5: 7–19.

Béteille, Andrè. 1995. 'Construction of Tribe.' *Times of India*, 19 June.

———. 1986. 'The Concept of Tribe with Special Reference to India.' *Journal of European Sociology* 27 (2): 297–318.

———. 1974. *Six Essays in Comparative Sociology.* New Delhi: Oxford University Press.
———. 1960. 'The Definition of Tribe', *Seminar* 14: 15–18.
Bose, Nirmal Kumar. 1971. *Tribal Life in India.* New Delhi: National Book Trust.
———. 1941. 'The Hindu Method of Tribal Absorption.' *Science and Culture* 77: 188–94.
Dube, S. C. (ed.). 1977. *Tribal Heritage in India.* Delhi: Vikas Publishing House.
Elwin, Verrier. 1944. *The Aboriginals.* Bombay: Oxford University Press.
Foster, George McClelland. 1953. 'What is Folk Culture?' *American Anthropologist* 55 (2): 159–73.
Fuchs, Stephen. 1973. *The Aboriginal Tribes of India.* Delhi: Macmillan India.
Godelier, Maurice. 1977. *Perspectives in Marxist Anthropology.* Translated by Robert Brain. Cambridge: Cambridge University Press.
Ghurye, G. S. 1963. *The Scheduled Tribes.* Bombay: Popular Prakashan.
Hardiman, David. 1987. *The Coming of the Devi: Adivasi Assertion in Western India.* New Delhi: Oxford University Press.
Kosambi, D. D. 1975. *The Culture and Civilization of Ancient India in Historical Outline.* Delhi: Vikas Publishing House.
Mandelbaum, David Goodman. 1970. *Society in India.* Berkeley: University of California Press.
Oommen, T. K. 1995. *Alien Concepts and South Asian Reality: Responses and Reformulations.* New Delhi: Sage Publications.
Pathy, Jaganath. 1992. 'The Idea of Tribe and the Indian Scene', in *Tribal Transformation in India,* Vol. 3, edited by Buddhadeb Chaudhuri. New Delhi: Inter-India Publications, pp. 43–54.
Ray, N. 1972. 'Introductory Address', in *Tribal Situation in India: Proceedings of a Seminar,* edited by K. Suresh Singh. Simla: Indian Institute of Advanced Study, pp. 3–24.
Redfield, Robert. 1956. *Peasant Society and Culture: An Anthropological Approach to Civilization.* Chicago: Chicago University Press.
Roy Burman, B. K. 1994. *Tribes in Perspective.* New Delhi: Mittal Publications.
———. 1983. 'Transformation of Tribes and Analogous Social Formations', *Economic and Political Weekly* 18 (27): 1172–74.
———. 1972. 'Tribal Demography: A Preliminary Appraisal', in *Tribal Situation in India: Proceedings of a Seminar,* edited by K. Suresh Singh. Simla: Indian Institute of Advanced Study, pp. 39–50.
Shanin, Teodor (ed.). 1973. *Peasants and Peasant Societies: Selected Readings.* Harmondsworth: Penguin.
Singh, K. Suresh. 1993. 'Marginalised Tribals.' *Seminar* 41: 212–18.
Sinha, Surajit. 1987. *Tribal Polities and State Systems in Pre-colonial Eastern and North Eastern India.* Calcutta: Centre for Studies in Social Sciences and K. P. Bagchi & Co.
———. 1965. 'Tribe–Caste and Tribe–Peasant Continua in Central India.' *Man in India* 45 (1): 57–83.
———. 1962. 'State Formation and Rajput Myth in Tribal Central India.' *Man in India* 42 (1): 35–80.
———. 1958. 'Tribal Culture of Peninsular India as a Dimension of Little Tradition in the Study of Indian Civilization: A Preliminary Statement.' *Journal of American Folklore* 71: 504–18.
Vidyarthi, L. P. and B. K. Rai. 1977. *The Tribal Culture of India.* Delhi: Concept Publishing Company.

Chapter 3

Tribes as Indigenous Peoples: Discourse and Adivasi Consciousness

The idea of characterising an 'indigenous people' in India has provoked of considerable debate in India today. This was not the case until a few years ago. Social workers, administrators, politicians, and even scholars used the term widely to refer to a certain category of people. They hardly felt any unease in using the native equivalent of this term, namely adivasi. Ghurye, who had some reservations about the use of such terms, employed the expression 'so-called aborigines'. Social workers, administrators, politicians, and scholars took the term, along with all its attendant prejudices, to the masses. That is how the notion of adivasi identity entered the consciousness of tribal people. The identity that was forced upon them from the outside precisely to mark their differences from the dominant community has now been internalized by the tribal people themselves. Not only has it become an important mark of social differentiation and identity assertion, it is also an important tool for the articulation of the demand for empowerment.

Inadequacy of the Term Tribe

The Anthropological Survey of India under the 'People of India' project identifies 461 tribal communities in the country. These are enumerated as 67,583,800 people, constituting 8.08 per cent of the total population as per the 1991 census. The scheduled tribes constituted 6.94 per cent and 7.85 per cent of the population in 1971 and 1981 respectively. The question of the identity of tribes in India is closely linked with administrative and political considerations. Hence, there have been increasing demands by groups and communities for their inclusion in the list of scheduled tribes under the Constitution of India. That partly explains the steady increase in the proportion of the scheduled tribe population in India, especially in the period between 1971 and 1981. Correspondingly, there has been more concern with the identification of tribes than with their definition. This assertion is not meant to convey the claim that lists have been drawn up without any prior conception of tribe whatsoever. That there did exist some idea is obvious from the criteria that were adopted, such as geographical isolation, simple technology, living conditions, general backwardness, animism, tribal language, and physical features. The problem, however, lay in the fact that the criteria were neither clearly formulated nor systematically applied. One set of criteria was used in one context and quite another set in another. The result is that the list includes groups and communities strikingly different from each other in respect of population size, level of technology, and other characteristics. Indian anthropologists have been acutely aware of a certain lack of fit between what their discipline defines as tribe and what they are obliged to describe as tribe. Yet they have continued with the use of this term.

The early ethnographers of India were not very clear in their minds about the distinction between caste and tribe in India. Eighteenth-century writings, for example, showed the synonymous use of the term tribe with caste. Later, it was even used in a cognate manner, as one can see in the use of the phrase 'castes and tribes of India' by Risley and many others in their writings. Efforts to make a distinction between the two terms began to be made after an attempt was made to collect detailed information about people for the census. The census officials were, however, far from clear about the criteria for the distinction. These criteria, however inadequate, are mentioned for the first time in the 1900 census. It defined tribes as those who practiced animism. In the subsequent censuses, animism was replaced by the category of tribal religion. Although the criterion was highly unsatisfactory, it continued to be used widely and extensively. It is only in the post-independence period that a more systematic effort was made to distinguish tribe from caste. Though the distinction between the terms was made in both colonial and postcolonial ethnography, the relation between the two was conceived differently in both periods. In colonial ethnography, the concern shown by British administrator–scholars was to mark off tribe from caste. Tribes were described as living in complete isolation from the rest of the population, and hence without any interaction with them. In contrast, the main concern in Indian ethnography has been to show the close interaction of tribes with the larger society or civilization. Both Ghurye and Bose, for example, stressed the nature of the interaction between tribes and the larger Hindu society and the ways in which tribes have been drawn into Hindu society. Sinha (1958) even goes to the extent of viewing tribes as a dimension of the Little Tradition, which cannot be understood adequately unless it is seen in relation to the Great Tradition. Under such a conception, tribes have come to be studied primarily in relation to the features and characteristics of the larger society, the focus being on how tribes are being absorbed into the larger society, the so-called mainstream, by becoming caste, peasant, class, and so on. With such a conceptualization, the identity of the tribal group or community is indeed put at stake. We are, led to such a proposition or conclusion because of the way in which tribes have been conceptualized in the anthropological literature and the frame of reference in which tribal society in India is primarily studied.

A tribe is seen primarily as a stage and a type of society. It is a society that lacks the positive traits of modern society and thus constitutes a simple, illiterate, and backward society. As these features change on account of education, modern occupations, new technology, etc., a tribal society is no longer considered tribal. If the transformation is in the direction of caste society, then it is described as having become a caste society. If the reference is to peasantization, then it is posited as a peasant society. If the general direction of transformation is towards social differentiation, then it is described as a differentiated or stratified society, and thus ceases to be a tribal society. In the process, it is forgotten that a tribe, besides being a stage and a type of society, is also a society similar to any other kind of society, such as the Oriya or the Bengali. But it is precisely this that comes to be denied on account of the changed situation. It is true that tribes are not of the same stage and type as Bengali or Oriya society. There is, then, something clumsy about the use of the term tribe to describe the Indian social reality.

The Concept of Indigenous Peoples

In the deliberations of the international agencies, the term indigenous people/population was used for the first time in 1957 in ILO Convention 107. It gained wide currency after 1993 with the declaration of that year as the International Year of the Indigenous People. The ILO convention 107 adopted in 1957 made a distinction between tribal and semi-tribal populations on the one hand and indigenous tribal populations on the other. The former were described as populations who were at a less advanced stage of development than those reached by other sections of national community and whose status is regulated wholly or partially by their own customs and traditions. The latter (indigenous) were, however, referred to as those categories of tribal or semi-tribal populations who traced their descent from the population which inhabited the country or a geographical region to which the country belongs at the time of conquest or colonization and which, irrespective of their legal, status live more in conformity with the social, economic and cultural institutions of that time than with institutions of the nation/country to which they belong (Daes 1996; ILO 1989; Roy Burman/undated). The ILO Convention 169 adopted in 1989 is substantively different in from those in Convention 107 in terms of conceptualization. To begin it speaks of with (1) tribal/indigenous peoples instead of tribal/indigenous populations and explicitly states their distinctiveness from other sections of the national community in the social, cultural and economic domains and (2) makes distinction between tribal peoples and peoples regarded as indigenous. In general, however, people, including scholars and activists, continue to use the term tribal along with indigenous.

Correspondingly, there has been a marked difference in the two conventions in their to general orientation towards tribal peoples. In 1957, the General Conference of the ILO adopted Convention concerning the protection and integration of indigenous and other tribal and semi-tribal populations in independent countries. The Convention framed general international standards for facilitating government actions towards protecting and promoting the progressive integration of these people into the respective national communities (ILO 1957). By 1985, the ILO felt the need to revise the convention on account of changes in attitudes and approaches towards these peoples world-wide. The ILO had earlier proposed integration as the desired objective but this was no longer seen as appropriate. This was so because the international organizations and an increasing number of governments were moving toward greater recognition of the rights of indigenous and tribal peoples to retain their specific identities and to participate fully in the planning and execution of the activities affecting their way of life. Accordingly, the ILO adopted a revised Convention (No 169) in 1989 after the expert committee appointed by the ILO gave its recommendation and the same was passed in consultation with other international bodies. While the ILO was moving in the direction of revising Convention 107 on tribal and indigenous peoples, the Working Group on Indigenous Populations, set up in 1982 by the Sub-commission on Prevention of Discrimination and Protection of Minorities of the Commission on Human Rights of the UNO, had already begun to employ the working definition of indigenous peoples developed in 1972 by Martinez Cobo, a Special Rapporteur of the UNO.

The definition was restated in 1986 by Cobo in his final report entitled 'Study of the Problem of Discrimination against Indigenous Populations', which reads as follows:

> Indigenous communities, peoples and nations are those which, having a historical continuity with pre-invasion and pre-colonial societies that developed on their territories, consider themselves distinct from other sectors of the societies now prevailing in those territories, or parts of them. They form at present

the non-dominant sector of society and are determined to preserve, develop and transmit to future generations their ancestral territories, and their ethnic identity, as the basis of their continued existence as peoples, in accordance with their own cultural patterns, social institutions and legal systems. (Daes 1996; Roy Burman undated).

The ILO Convention referred to above and the Working Group on Indigenous Populations set up by the Human Rights Commission of the UNO (Cobo Report) speak of the indigenous peoples/populations as follows: They are those tribal and semi-tribal peoples/populations regarded as having their descent from the peoples/populations which inhabited the country or the geographical region to which the country belongs, at the time of conquest or colonization. They are, in addition, also those who, irrespective of their legal status, live more in conformity with their own social, economic and cultural institutions than with the institution of the nation to which they belong. Thus, there are two aspects central to the conceptualization of indigenous peoples. First, the indigenous are those people who lived in the country to which they belong before colonization or conquest by people from outside the country or the geographical region. Secondly, that the people govern their life more in terms of their own social, economic and the cultural institutions than the laws applicable to the society or the country at large. What is important here is that the notion of indigenous peoples, despite sharing attributes in common with the peoples described in ILO Convention 169 as the tribal peoples (in ILO Convention 107 as tribal and semi-tribal populations) is different from the latter in the sense that indigenous peoples are invariably marked out as distinct international entities. That is, the indigenous are invariably victims of conquest and colonization from outside. Furthermore, outsiders are easily identifiable. Being the descendants of the peoples who lived in their territory before conquest or colonization is the most important criterion towards definition of indigenous peoples, but it is not the only one. Other criteria include non-dominance or colonial situation, and living more in conformity with their own social, economic and cultural customs and traditions than the institutions of the country of which they now are part. In short, the context of the discourse on indigenous peoples, initiated by the ILO and later accepted by the UNO basically hinged on the twin concepts of 'need right' and 'power right' of a certain social category of people all over the world. This category of people was progressively being marginalized and dispossessed from their sources of livelihood and was vulnerable to cultural shock and loss of their collective identity.

Tribes as Indigenous

The term indigenous people, though used more recently at the international level, has been in use in India for a long time. Social workers, missionaries, and political activists have been using the term adivasi, the indigenous term for indigenous people, freely to refer to tribal people since the beginning of the twentieth century. The term in conjunction with other related terms, such as aborigines and autochthones, has also been used extensively by scholars and administrators. It was used mainly as a mark of identification and differentiation, that is, to mark out a group of people different in physical features, language, religion, customs, social organization, etc. Even Ghurye who otherwise talks of tribes as backward Hindus, refers to them as aborigines. He writes:

When the history of [the] internal movements of peoples is not known, it is utterly unscientific to regard some tribe or the other as the original owner of the soil. It is possible to contend that even if the tribes are not aborigines of the exact area they now occupy, they are the autochthones of India and to that extent they may be called the aborigines. (1963: 12).

Scholars used the term to refer to these groups without any apparent sense of unease. No effort whatsoever was made to dispel the myth associated with the term. Scholars, social workers, political activists, and administrators popularized the term along with all its concomitant prejudices (Sengupta 1988). It came to be used widely to refer to tribal people. Its use was hardly questioned, let alone debated. So long as it did not assume political dimensions, it remained an accepted term for the description and designation of a certain category of people.

It is only with the internationalization of the context, and the rights and privileges associated with it, that the use of the term indigenous has come to be critically examined, or even challenged, in the Indian context. The sense in which the term was used earlier and the sense in which it is used today are definitely not identical, though they overlap in some sense. Today, aspects of marginalization are built into the definition of indigenous people. Only those people who were subjected to domination and subjugation have come to constitute the category of indigenous people. Yet the use of the term adivasi (indigenous) to designate a certain category of people clearly reveals that these aspects were not altogether lost sight of. Even earlier the term had been used to describe people who were backward and cut off from the mainstream of civilization. The basic mark of differentiation was between those who were part of civilization and those who were not. Hence, the use of the term adivasi to describe tribal people seems to have some validity, even in the sense of marginalization. The requirement of historical antiquity may have been a more distant criterion, but the most immediate and proximate criterion seemed to be the fact that they were not part of civilization. In a certain sense, the aspect of marginalization was taken into consideration when designating a group adivasi. This is clear when we take into account other aspects of Indian society. The coming of the Aryans has invariably been taken as the decisive historical factor in determining the original people of India. Yet not all original peoples have been called indigenous peoples. The groups speaking languages belonging to the Dravidian linguistic stock have no doubt been considered the inhabitants of India since before the coming of the Aryans. Yet they have never been described as the indigenous people, mainly because they do not constitute marginalized groups. The Government of India had, in fact, expressed no objection to the use of the term when the matter was considered by the ILO convention in 1957 and was tied to Covenant No. 107. This was because the term at the time had not come to be associated with issues such as minority empowerment and rights. Rather, it had pointed to the need for integrating the indigenous and tribal people into the larger social and political system. The focus shifted from integration to rights and empowerment by 1989. No sooner had the focus shifted than the argument was made that a category such as indigenous people did not hold up well in the Indian context.

Arguments Against

Much of the discussion on the indigenous people/status of tribes in India has centred on the complex historical processes of the movement of various peoples and their settlement in the subcontinent. It is said that unlike in the Americas, Australia, and New Zealand, where identification of indigenous people is easier on account of the recent history of conquest, immigration, and colonization, the same is not the case in India. Rather, in India, movements of peoples of different languages, races, cultures, and religions date back over millennia. Even groups described as tribes have not been outside of this process. Given this, how far back should one go in history to determine who is a native and who an immigrant? Any demarcation is bound to be arbitrary and hence extremely contentious. This has indeed been the case, as we can see from the discussion below. It is also maintained that the communities described as tribes have been living in close interaction with non-tribal people for centuries, leading to a great deal of acculturation and even assimilation into the larger Hindu society. The Indian experience, it is stated, is different from that of the New World, which was marked by conquest, subjugation, and even genocide. It is hence argued that it is not only the point of departure that is problematic but also the nature of the Indian experience.

In India, the term indigenous people has generally come to be associated with the people described as tribes. It is assumed that they were the original settlers of what geographically constitutes India today, or at least were the people who inhabited the region before the coming of the more dominant sections of Indian society, namely the Aryans. They are said to belong to racial stocks other than the Aryan and to speak a variety of dialects belonging presumably to two major linguistic families, namely the Dravidian and the Austric. The plausibility of groups speaking the Tibeto-Burman language is not altogether ruled out from the purview of the status of the original inhabitants in India. Ray (1973: 124–25) writes:

> The communities of people of today whom the anthropologists call tribals, happen to be the indigenous, autochthonous (adivasis, adimjati) people of the land, in the sense that they had long been settled in different parts of the country before the Aryan-speaking peoples penetrated India to settle down first, in the Kabul and Indus valleys and then[,] within a millennium and [a] half, to spread out in slow stages, over large areas of the country and push their way of life and civilization over practically the entire area of the country along the plains and the river valleys.

The questions of central importance here are: (1) whether groups designated as tribes were natives of India and of non-tribe immigrants; and (2) if they were not natives, whether their settlement was prior to that of the arrival of the major social groups, namely the Aryans. Most scholars are of the view that the tribes could hardly make a legitimate claim to being native to India. They cite the observations of certain scholars, however conflicting these may be, in support of their position. Hutton (1931), for example, is of the view that only the Negritos may be considered the original inhabitants of India, although they do not have a marked presence now. He considers groups belonging to the Austric, Dravidian, and other categories as much outsiders as the Aryans. Guha (1944) is also cited for making a similar observation in the context of Austric-speaking people. But more authoritative sources on the basis of which such a claim has been questioned are the traditions of the tribes themselves. Dube (1977: 2) writes:

> It is difficult to speak of 'original' inhabitants, for tribal traditions themselves make repeated mention of [the] migration of their ancestors. There is considerable evidence to suggest that several groups were pushed out of the areas where they were first settled and had to seek shelter elsewhere. And there are several groups, now absorbed in Hindu society, which can make an equally tenable claim to being original or, at any rate, very old inhabitants.

Two substantive points have been made here. The first is whether one can speak of tribes as the original people, especially in view of their migratory movements. This is indeed an important point and cannot be brushed aside. The second makes reference to the claims that may be made by groups that have been absorbed into Hindu society. The latter suffers from two flaws. First, it is purely hypothetical. Second, the groups lose the right to make such a claim by virtue of their decision to be absorbed into dominant Hindu society.

If the issue of tribes as natives is questionable, is this also the case with the second issue? That is, are all tribal groups those who settled the territory before the coming of the Aryans, the assumed cut-off point for determining the status of indigenous people in India? Whereas this is more or less the case, this cannot be said with complete certainty for all the groups described as tribes in India. It is said that there are tribes in India, especially in the north-east, whose settlement in the territories where they live today is an even later development than the settlement of many non-tribes in other parts of India. The Nagas, for example, are said to have come to India around the middle of the first millennium BC, first to Tibet and later to the territory where they live now, a period later than the coming of the Aryans. The Mizos are said to have settled in the territory where they live now only in the sixteenth century. The settlement of the Kukis is considered an even later development than the settlement of the Mizos. In contrast to this, non-tribal groups like the Bengalis, the Gujaratis, and the Oriyas have a much longer history of settlement than do the tribes referred to above. It is indeed problematic to say, then, that all tribal people in India were earlier settlers than the Aryans, and that, therefore, tribes are indigenous and non-tribes are non-indigenous. Yet in the context of the argument made, there is a need to make a distinction between settlement in the context of a country (India being the reference point here) as a whole and settlement within its parts or regions. In the discourse on indigenous people, these two aspects are either ignored, mixed or even interchanged. Correspondingly, the argument either in defence or against the proposition is valid at one level. For example, at the local level, it is often used to substantiate the argument at another level, namely the country as a whole. The Santhals may have settled in the territory where they live now, namely the Santhal Parganas or areas adjacent to it, at the beginning of the nineteenth century. They may have even settled there later than the Bengalis. But that in no way rules out the fact that their settlement in India is prior to that of the Aryans, including the Bengalis. Conversely, the settlement of the Mizos in the country called India may have been a later development than the settlement of the Gujaratis or the Bengalis, but the fact remains that they were the original settlers of the place where they now live.

It needs to be mentioned here that tribal groups in India do not consist solely of groups speaking the Dravidian and Austro-Asiatic languages. A very large number of tribal groups, in fact, belong to the Tibeto-Burman-language-speaking group, many of whom can hardly be considered indigenous if the arrival of the Aryans is taken as the cut-off mark to decide who is indigenous and who is not. To restrict the term indigenous to only those groups that entered the subcontinent prior to the arrival of the Indo-Aryans would mean excluding many tribal groups of the Tibeto-Burman-language-speaking family from the status of indigenous people. There are also tribal groups like the Bhils who speak languages of the Indo-Aryan family. This poses the problem of their identification as indigenous people. Yet it is generally held that tribal groups have adopted the languages they speak through the processes of interaction and acculturation with Indo-Aryan-speaking groups. In terms of their cultural and physical features, they are, however, still considered different from the Indo-Aryan population (Hermanns 1957: 23). The fact that they speak a language belonging to the Indo-Aryan group is attributed to the acculturation process resulting from contact between the Bhils and the Aryans. What this means is that people identified and described as tribals should not necessarily be treated as indigenous and that there are tribal groups that could be treated as indigenous and others that could not. In contrast, many groups and communities, especially those belonging to the Dravidian-language-speaking group such as Tamilians, the Telugus, and the Malayalis, could stake a claim to being indigenous by virtue of the fact that they were inhabitants of India prior to the coming of the Aryans. They are, however, not recognized as tribals, and share few attributes in common with the tribals, who stand dispossessed, exploited, and marginalized. Such is not the case with the groups referred to above. Rather, they constitute a part of the dominant national community. In terms of other criteria for indigenous people, status, namely marginalized status, loss of control over resources, etc., they can hardly be considered eligible. The congruence between the term and the concept on the basis of which tribal activists defend the application of the term does not stand valid in all situations. There is another point that needs to be made in this context. The description of groups and communities as indigenous on the basis of the language usage raises another problem. There are many tribal groups that speak languages of the Indo-Aryan group. This poses the problem of identification of the groups concerned as indigenous people. These groups may have adopted these languages through the processes of interaction and acculturation with groups already speaking these languages. Yet this does not at all contradict the other fact that the present aborigines are completely different in terms of their ethnicity and culture from modern Aryans. Rather, it confirms the clear distinctions that have been made between them for centuries. There is still another ground on which the indigenous claim is contested in the Indian context. It is generally held that Indian society is made up of a number of castes and groups, and that many of these are the result of a process of fusion of various groups and communities, including tribes. In view of this, it becomes extremely difficult to reconcile the claim that a segment of one community can be identified as indigenous and another as non-indigenous.

In Defence

The extension of the term is, however, strongly defended by activists and scholars, both tribal and non-tribal. This is done not so much on the basis of original settlement as on some other consideration. They, of course, trace the history of tribals in India much before the coming of the Aryan people. They also argue that tribals, whose ancestors lived here for thousands of years prior to the Aryan invasion—a fact that can hardly be disputed—should be considered indigenous people, so that certain positive international instruments can be made applicable to these marginalized and deprived social groups. But the case is made more forcefully on other grounds. Pathy (1992a: 8), for example, writes that tribals have been victims of conquest and colonization, and hence share all the attributes of colonized people, such as ethnic identity, loss of control over customary territorial resources, cultural annihilation, and powerlessness. He makes this case despite recognizing the fact that the claim of original settlement in a territory is problematic and unreasonable. But the issues of colonization and colonized status remain far from being resolved. First, we do not have detailed and well-researched historical evidence on the nature of the relations or encounters between groups designated as indigenous and other social groups for the period preceding the coming of the British, which would allow us to draw firm conclusions about the nature of relations between the two types of social groups. In general, the relation between tribes and non-tribes has been described as one of mutual coexistence rather than one of subjugation and domination, at least until the advent of British rule. This is the other important ground on which the term indigenous people has been adopted in the Indian context. It is said that most studies on the history of Indian civilization depict the growth and expansion of Hindu society as a prolonged and complex process of assimilation. The nature of the interaction between tribals and non-tribals has been broadly described as one of peaceful coexistence rather than one of conquest and subjugation (Bose 1941; Béteille 1998: 189). This has also been endorsed by a scholar who otherwise talks of an aggressive absorption of tribals into Hindu society with the onset of colonial rule (Desai 1977: 24). Pathy (1992b: 51) speaks elsewhere of a symbiotic relationship between tribals and non-tribals rather that a relationship based on colonization and conquest. He writes, 'Majority of the so-called tribals of India had developed [a] class structure over a long period of time and therefore had interactions with the other communities. Up to the time of colonialism, it was largely not a relationship of domination and subjugation.' Not only is the question of original settlement in a territory problematic, so are the questions of colonization and subjugation, as one can see from Pathy's observation. This means that the question of indigenous people could only be raised from the period of the arrival of the British in India and the subsequent processes of colonization and subjugation. If these processes are taken as the point of departure for the determination of indigenous people status, then the issue of original settlement so central to the notion of indigenous people, in fact, becomes redundant.

In fact, the whole exercise of the identification of groups and communities as tribes during the colonial period was to a great extent contingent upon the nature of the relations that the tribes had with the larger society. This factor seems to have been accepted without much deliberation or debate as an important consideration in the identification of tribes in India. By and large, tribes were considered as those groups that were outside civilization. This means that they not only remained outside the politico-administrative structure of the larger society or the kingdom, but also outside the general social organization and world-view of the larger society. In short, they continued to be distinct because they escaped colonization and subjugation. Where people were subjugated, they became a part of the larger social organization and failed to maintain their distinctiveness. That they lived on hills and plateaus or in forests, and survived on hunting and food-gathering or practiced slash-and-burn agriculture, followed from being out of the civilization complex. Emphasizing this point, Béteille (1986: 316) writes that where tribe and civilization coexist, as in India and in the Islamic world, being a tribe has been more a matter of remaining outside of state and civilization, whether by choice or necessity, than of reaching a definite stage in the evolutionary advance from the simple to the complex. The Indian practice of regarding tribes as a large assortment of communities, differing widely in size, mode of livelihood, and social organization, cannot, therefore, be dismissed as anomalous. They are all tribes because they all stood more or less outside of Hindu civilization, and not because they were all at exactly at the same stage of evolution. In short, they are described as tribes, and hence even as indigenous people, because they escaped the processes of colonization and subjugation.

The use of the term indigenous people to refer to tribal people is defended on yet another ground. It is argued by some scholars that unlike antagonists. Who tend to take note of only historical realism, we have also to take note of critical realism. They state that irrespective of the place and time of origin or occupation of the present habitat in India, there are certain communities which until recently maintained practically autogenous sources of legitimization of cultural and social processes, which followed an ideology of a self-regulated economy, and which had only marginal interaction with external political structures. Their indigenous identity cannot be brushed aside by the juxtaposition of non-meaningful occurrences in space and time in systemic terms (Roy Burman 1992: 24). Such a claim is difficult to establish today. Not only had tribal communities been brought under uniform administrative and legal structures under the British, they had also been drawn into the politico-economic processes of the larger society, especially in the post-independence period. Very few tribes have managed to escape such processes. The empirical reality is far too complex. It is not at all surprising, then, that elsewhere Roy Burman (1983: 1172–74) writes, 'Even many of those with the simplest technology were integrated with the wider society.' In short, the use of the term indigenous to describe tribal people in India is fraught with many difficulties. It is less reflective of the empirical reality, being more in the nature of a political construction.

Though the question of indigenous people has been discussed generally in the context of India as a whole, the discussion has taken place in relation to different regions and territories within the country. There have been so many migrations in and out of the subcontinent over the centuries that no particular jati can have genuine grounds for making such a claim (Béteille 1998: 189; Hardiman 1987: 15–16). In any discussion of indigenous people in India, it is important that we do not mix problems obtaining at two levels. Often, the problem at the level of the country is used to make a case against tribes being indigenous at the regional or local levels, and vice versa. Posing the question of tribes as indigenous people in relation to territories within the country gives rise to problems of a somewhat different nature. The movement of people belonging to different races, ethnicities, and linguistic groups, including those described as tribals, from one place to another is something that has been taking place in India for centuries. Groups that may be indigenous with respect to the country as a whole may not be indigenous with respect to their settlement in a given territory. It may also happen, as is usually the case, that the same group may be indigenous and not indigenous at the same time. The Oraons, the Mundas, and many other tribes living in Jharkhand, for example, may have a legitimate claim to be called indigenous people in respect of their settlement in the country called India prior to the arrival of the Aryans, or even in respect of their settlement in Jharkhand, but it is not certain if they can claim to be indigenous in Assam and Bengal, areas into which they moved during the course of the last century or so. Indeed, their claim to being indigenous is strongly contested in these places. Nowhere is this more true than in Assam where the migrant tribals' claim to be the indigenous people is disputed by tribal communities such as the Bodos and the Mishings, who have a much longer history of settlement in the region than the tribals from Bihar, Orissa, Madhya Pradesh, etc. If, however, one takes India as a whole, then these migrant groups have a much longer history of settlement than the tribes in the north-eastern region. The tribals from this region have also moved in large numbers to the Andaman Islands and have settled there as cultivators after reclaiming land from the forest, and in the process dispossessed the native Jarvas from their territories. Thus, the Oraons, the Mundas, and others are indigenous in some context, that is, in the context of their claim to have lived in India before the Aryan invasion, and also in the context of their being the first settlers of Jharkhand. Their claim to be indigenous, however, is disputed in places outside the Jharkhand region.

Adivasi Consciousness

In the context of India as a whole, then, the determination of indigenous peoples is indeed problematic. The problem, as mentioned earlier, is said to emanate from the fact that population movements and experiences in India have been different from those in the New World. There is another difference, which has not been adequately highlighted. It is true that the movement of population even in the New World has been similar to the one in India—of different races characterized by different languages, origins, religions, cultures, etc. In the New World, however, these groups did not come to establish or have special relations with a given territory or region in the course of their movement. In India, on the other hand, population movements were somewhat different. Here different communities came to develop a distinct and definite association with certain territories in the course of their movement. Whether those especially associated with a given territory are indigenous to the territory or land in which they live is a question that will always be contested. However, what remains uncontested is the fact that they have developed special relations with the territory in question. The communities in question considered these territories as their own against the claims of other communities. They considered themselves as having prior and preferential, if not exclusive, rights over the territory in which they lived, either on account of their prior historical settlement or their numerical dominance. Following this, they sought to promote and protect the interest and welfare of their community and to demand special rights and privileges. These aspirations led to the desire to have a state of their own. Having articulated this demand, they sought to promote the interests of their community through state patronage. The people living in their respective territories have thus come to see the states to which they belong as the culmination of the yearning of the residents of a particular territory to have a homeland of their own. This indicates recognition of the fact, implicit though it may be, that certain people have prior rights over others in the territory that they occupy. This is almost like saying that they are the original inhabitants of the territory that they inhabit. It is, therefore, not a coincidence that the dominant communities hardly feel the need to articulate issues along the lines of indigenous versus non-indigenous people. By virtue of the fact that they have states of their own, and hence territories too, they take the matter for granted. The issue surfaces only when they feel threatened by the movement of people from outside the community. They fear either the increasing numbers of outsiders or the loss of control of power, both economic and political. Nowhere is this close identification with land or territory more clearly manifested than in the son-of-the-soil theory, raised from time to time by the dominant regional communities in India. In short, people speaking different languages, possessing different physical features, following different cultures, and observing different modes of social organization, identify and relate themselves in a special way with a given territory or region. Attempts to theorize such developments in India have been made. In doing so, scholars have invariably made use of concepts such as nation and nationality. The element of territoriality assumes a central place in both these concepts.

Paradoxically, whereas such privileges and rights are freely recognized with respect to these communities, the same is denied to tribal communities in India. In the process, they are being dispossessed of their control over resources (land, forest, water, minerals) in their own territory and are being subjected increasingly to misery, injustice, and exploitation. If their status as the indigenous people of India is problematic—and the problem indeed is both empirical and conceptual—the least that the dominant regional communities could do is to recognize the rights and privileges of these people in the territories and regions they inhabit.

In the absence of such powers and rights, a new form of identity, namely the identity of adivasis or indigenous people, is crystallizing among tribes in different parts of India. The term, adopted mainly as a point of reference or description, has today become an important marker of identity articulation and assertion. As noted earlier, the designation or description of tribes as indigenous people was not the result of self-identification or self-description by the tribal people themselves. It was not part of the attempts at positive self-identification and self-evaluation by the tribes. Rather, the term was used by outsiders to describe the tribes. The identity imposed by outsiders has been accepted and internalized by the tribes. It is an important marker of identity and consciousness, which evokes a sense of self-esteem and pride rather than a sense of inferiority and shame, feelings that often accompany terms like tribe and tribal. Tribal people now use it to identify and define themselves. It is on the basis of adivasi identity that tribes are increasingly differentiating themselves from non-tribes, at least at the grassroots level. The declaration of 1993 as the International Year of the World's Indigenous People has sharpened this identity, for this identity has now come to carry certain rights and privileges with it.

Further, the identity being expressed through the term adivasi is indeed an expanded one, cutting across tribes bearing different names and speaking different languages or dialects. It also goes beyond the groups and communities, or a part thereof, listed in the Constitution. It should be noted that there is an important gap in the sense in which the term tribe is used and understood by tribal people and the sense in which it is understood by others, especially administrators, lawyers, and academicians. For the latter, communities are tribes only if they are listed in the Constitution. Tribes, on the other hand, do not view themselves in the sense of constituting a politico-administrative category. Rather, they view themselves in the sense of belonging to the same community, irrespective of whether a group, or a segment of it, is listed or not listed in the Constitution. By virtue of this bond of emotion, they are also adivasis or indigenous people, though the Constitution does not recognize them as tribes.

In many parts of the country, the category had taken on political overtones. In Chhotanagpur, for example, as early as 1939, it assumed the form of a demand for a separate state for the adivasis. The Adivasi Mahasabha, the organization formed to spearhead this movement, drew its inspiration from being described as the representative of the original people. Such political overtones are not confined to Chhotanagpur alone. The movement has spread to other parts of India as well, particularly the tribal regions of central and western India, such as Madhya Pradesh, Gujarat, and Rajasthan. This has led to the emergence of many organizations at the local and regional levels articulating tribal demands along the lines of the status of indigenous people. Some organizations are even actively articulating the demands of the indigenous people at the national and international levels.

Adivasi consciousness and the demand for indigenous people status are not so much about the claim to being the original inhabitants of India as about the fact that the tribals have no power whatsoever over the resources (land, forest, river, minerals) in the territory in which they live. This despite the fact that they are prior inhabitants of India in relation to other groups. Adivasi consciousness and the articulation of political demands are basically expressions of a yearning for a special relation with the territory in which they live. It is the same kind of yearning that various dominant communities in India articulated in the period before and after independence. This is seen clearly if one maps the tribes and regions where such a sentiment is strongly articulated. This identity is more strongly articulated in central, western, and southern India than in north-eastern India. The reason why this articulation of tribal identity is weak in the north-east is because people here exercise some power over the territories in which they live. The scenario is just the opposite in other parts of tribal India. This indicates that the assertion of such an identity is stronger where there is a greater degree of marginalization and powerlessness.

There are, however, differences in the way that this new identity is being conceptualized at different levels. Social workers, administrators, scholars, and social scientists have generally used the term in the sense of only the original inhabitants. Adivasi ideologues too primarily use this term in the sense of original settlement, as this was the sense in which outsiders described tribal people. The aspect of marginalized status that evolved in the course of history has now been added to the meaning of the term. Others, especially radical scholars, see it only in relation to a particular historical development, namely the subjugation during the nineteenth century of various communities that before the colonial period had been free, or at least relatively free, of the control of outside states. This process, it is argued, was accompanied by an influx of traders, moneylenders, and landlords, who established themselves under the protection of the colonial authorities and took advantage of the new judicial system to deprive the adivasis of large tracts of land. Adivasis are hence defined as groups that shared a common fate in the nineteenth century, and which, as a result of this experience, have evolved a collective identity, that is, an identity of being adivasis. This is not meant to imply that adivasis are the original inhabitants (Hardiman 1987: 15–16). The use of the term adivasi in the sense employed by radical scholars hardly takes note of the sense in which people, either outsiders or the tribals themselves, use it. Of course, not all groups that shared a common fate in the nineteenth century identify themselves as adivasis. Sections of the caste Hindu population, however deprived it may have been, invariably avoided being called adivasi for the term implied a loss of status.

References

Béteille, Andrè. 1986. 'The Concept of Tribe with Special Reference to India.' *European Journal of Sociology* 27 (2): 297–318.

———. 1998. 'The Idea of Indigenous People.' *Current Anthropology* 39 (2): 187–91.

Bose, N. K. 1941. 'The Hindu Method of Tribal Absorption.' *Science and Culture* 7.

Daes, Erica-Irene A. 1996. 'Standard-setting Activities: Evolution of Standards Concerning the Rights of Indigenous People.' Working Paper by the Chairperson. United Nations Economic and Social Council, UN Document E/C.N.4/Sub. 2/AC.4/1996/2.

Desai, A. R. 1977. 'Tribes in Transition', in Romila Thapar (ed.), *Tribe, Caste, and Religion*, New Delhi: Macmillan, pp. 15–28.

Dube, S. C. (ed.). 1977. *Tribal Heritage of India*. New Delhi: Vikas Publishing House.

Ghurye, G. S. 1963. *The Scheduled Tribes*. Bombay: Popular Prakashan.

Hardiman, David. 1987. *The Coming of the Devi: Adivasi Assertion in Western India*. New Delhi: Oxford University Press.

Hermanns, Matthias. 1957. *Hinduism and Tribal Culture: An Anthropological Verdict on the Niyogi Report.* Madhya Pradesh (India). Bombay: Christian Missionary Activities Enquiry Committee. K. L. Fernandes [foreword 1957].

ILO. 1957. 'Indigenous and Tribal Populations Convention, 1957 (N.107).' Convention Concerning the Protection and Integration of Indigenous and Other Tribal and Semi-Tribal Populations in Independent Countries. Geneva: ILO.

———. 1989. 'Indigenous and Tribal Populations Convention, 1989 (N.169).' Convention Concerning Indigenous Tribal Peoples in Independent Countries. Geneva: ILO.

Pathy, J. 1992a. 'The Idea of a Tribe and the Indian Scene', in *Tribal Transformation in India,* Vol. 3, edited by Buddhadeb Chaudhuri. New Delhi: Inter-India Publications.

———. 1992b. 'What is Tribe? What is Indigenous? Turn the Tables Toward the Metaphor of Social Justice.' *Samta* 1: 6–12.

Ray, Niharranjan. 1973. *Nationalism in India: An Historical Analysis of Its Stresses and Strains.* Aligarh: Aligarh Muslim University.

Roy Burman, B. K. 1992. 'The Indigenous Peoples and Their Quest for Justice', in *Tribal Transformation in India,* Vol. 3, edited by Buddhadeb Chaudhuri. New Delhi: Inter-India Publications.

———. 1983. 'Transformation of Tribes and Analogous Social Formation.' *Economic and Political Weekly* 18 (27): 1172–4.

———.n.d. 'Indigenous and Tribal Peoples and the UN and International Agencies.' New Delhi: Rajiv Gandhi Foundation (Mimeograph).

Sengupta, N. 1988. 'Reappraising Tribal Movement.' *Economic and Political Weekly*, 23 (20): 1003–1005.

Sinha, Surajit. 1958. 'Tribal Cultures of Peninsular India as a Dimension of Little Tradition in the Study of Indian Civilization: A Preliminary Statement.' *Journal of American Folklore* 71: 504–518.

Vidyarthi, L. P. and B. K. Rai. 1985. *Tribal Culture of India*. New Delhi: Concept Publishing Company.

Chapter 4

Tribes and Citizenship: Making Sense of Citizenship Rights

The broad assumption underlying the concept of tribes in India is that these are groups and communities located on the fringes of Indian civilization. They are seen as having been drawn and absorbed into Indian civilization throughout history through a variety of social and political processes. Yet the process of incorporation was far from complete when India became independent in 1947. In the post-independence period, the position of the intelligentsia in general and that of the Indian state in particular with respect to tribes has been one of integrating them into the larger Indian society. This was intended to enable tribes, in the phraseology of citizenship rights, to share in full in the social heritage and life of a civilized society. To this end, tribes, like other sections of the Indian population, were given citizenship rights after India declared itself a sovereign democratic republic. This chapter examines the tribes in India in the light of citizenship rights. What have citizenship rights so far meant to tribal people? To what extent have they been able to take advantage of such rights? Have these rights made any difference to their lives? Has there been any substantive contribution to efforts aimed at increasing the tribes' share in the social heritage and life of a civilized society?

Citizenship Rights

Citizenship entitles an individual to full membership of a community. It confers an array of rights and obligations upon them. In this sense, the status of an individual as a citizen is contrary to the general notion of status, which is invariably associated with the notion of hierarchy and inequality (Béteille 1996). Citizenship rights are the hallmark of a democratic political system. Citizenship has three components:—civil, political, and social rights.

Civil rights guarantee individual freedom, such as the right to individual liberty, right to freedom of speech, thought, and faith, right to own and transfer property, and right to justice. The right to justice guarantees equal treatment before the law in defending and asserting one's rights and the right to be subjected to the same processes of law as everyone else.

Political rights consist of the right to stand for election to a representative body vested with political authority, such as the parliament and its counterparts at the regional and local levels, as well as the right to elect one's political representatives.

Social rights consist of the right to economic welfare and security, the right to participate to the fullest in the social heritage, which means the right to live the life of a civilized being, according to the standards prevailing in the society (Marshall 1977: 78–91). Heater describes the nature of this citizenship right as emerging from 'the Liberal tradition' as against the other tradition, which he calls 'the Civic Republican tradition'. The former places emphasis on rights, the latter on duties (Heater 1999: 4).

Situating Citizenship Rights in the Indian Context

Until India gained independence, people had no citizenship status. They had no civil rights for they did not have the right to personal freedom. They could not take up the occupation of their choice nor were they subject to the same process of law. They also did not have the right to freedom of speech, expression, and faith. Nor did they have political rights in the sense of the right to the exercise of political authority and the right to elect their political representatives. However, they did enjoy a modicum of social rights in the sense of social security. This came from the social arrangement of caste. They enjoyed security within the overall framework of caste hierarchy and inequality. This meant that social rights derived from status, the hallmarks of which were class and hierarchy. Further, the security so enjoyed was localized. Under the British, some rights were granted to the people at large not so much as a part of the entitlement to rights but more so as a corollary or end product of colonial administration. For example, the British introduced uniform civil law and administration. In the process, people irrespective of their caste and creed came to be governed and administered by the same law, unlike the situation in traditional Indian society, which was governed by laws contingent on the individual's caste position. This was the origin of the right to justice in Indian society.

Indeed, many aspects of citizenship rights that people came to enjoy, albeit in a very limited sense, came not as part of the drive to equality, as was the case in Europe, but as part of the general process of societal development under the colonial regime. The processes of industrialization and urbanization and the introduction of modern education loosened the fetters of traditional Indian society and widened the scope for social mobility. Social mobility posed a challenge to the lack of freedom and inequality inherent in the whole notion of birth-based status. It permitted freedom to some degree, and to that extent it was a move towards equality, but of a certain kind. People also came to enjoy some other rights, but these were won through struggle during the course of the national movement. The right to freedom of speech and expression, although limited, was one among them. The rights so won were more in the civil domain rather than in the political and social domains. Some political rights were won during the colonial period, but they were limited in scope. These rights were restricted only to people of certain social and economic standing. For instance, property owners could acquire political rights during the colonial period.

The scene changed dramatically after independence. India adopted a Constitution that committed itself to extending not only civil and political rights but also social rights to all it people. The spirit underlying these rights was based on an ideology of social justice, liberty, equality, and fraternity. In doing so, the framers of the Constitution made a radical departure from the tradition based on an ideology of inequality and hierarchy, on the one hand, and lack of freedom and social justice, on the other hand. To use Béteille's formulation, India moved from a harmonic to a disharmonic system with the adoption of the Indian Constitution. That is, India committed itself to the ideology of equality, but the social system was characterized by hierarchy, inequality, and non-egalitarianism both in terms of ideology and practice (Béteille 1974). For millions of people living under hierarchy, inequality, and lack of freedom, this was indeed a landmark development in Indian history.

However, even more interesting is that all these rights, namely civil, political, and social rights, came at the same time in the case of India. In Europe, these rights came, or rather were won, at different times. The origins of civil rights can be traced to the eighteenth, those of political rights to the nineteenth, and those of social rights to the twentieth centuries. The granting of these rights was built successively on the foundation laid by the previous rights (Marshall 1977: 78–91). This was not the case in India. The preamble to the Constitution of India, for example, promises to secure to all citizens justice—social, economic, and political—liberty of thought, expression, faith, and worship, equality of status and opportunity, and fraternity assuring the dignity of the individual and the unity and integrity of the nation (Bakshi 2000: 1). It is important to note that the Indian Constitution talks of the nation. It is doubtful if political scientists and sociologists would talk in the same vein. At the same time, it is also important to note that the rights enshrined in the Constitution had not emerged out of strong movements and mass mobilizations against the iniquitous caste system. The freedom movement, against the backdrop of which the role and importance of citizenship rights in post-independence India should be understood, did not produce very strong movements of mass mobilization against social inequality in general, and against caste hierarchy in particular. This is not to deny the long history of protest movements against the iniquity of the caste system. These had taken the form of either reforms initiated by enlightened upper-caste social reformers or, more specifically, movements and social mobilizations launched by people placed lower on the caste hierarchy.

Citizenship Rights and Tribes

Indian society is marked by plurality. The major pluralities are those of language, caste, religion, and tribe. The status of citizenship does not make a distinction on the basis of these pluralities. The status bestowed assumes crucial importance in the context of marginalized groups and communities. What significance do they attach to such a development? How have they made sense of it? Do these rights mean the same thing for people belonging to different castes and classes? Do they hold the same importance for them all? These seem to me interesting questions. It is equally important to examine the extent to which citizens have been able to exercise the rights conferred on them. After all, marginalized groups are far from homogeneous. They are situated in different contexts and have varying experiences, interests, and aspirations. Hence, it is not possible to examine the issue of citizenship in the context of marginalized groups as such. This chapter attempts to examine citizenship rights in the context of tribes in India. Tribes, as we all know, have had experiences quite different from those of groups in caste society and those on its fringes.

It may be kept in mind that citizenship rights are individual rights. Tribes are communities. Individuals have, at the same time, no social existence outside of their community or society. Rather they can only be understood in and through the society in which they live. In a pluralistic society like India, individuals are invariably members of a given community. Not only that but also individuals as citizens have the right to work for the welfare and development of their community as well as for safeguarding and protecting its identity. This is how dominant communities, whether linguistic or religious or other, have consolidated their position in Indian society. All these rights are, of course, subject to the rights of other individuals as citizens. In reading tribes as citizens, therefore, I am treating tribes both as citizens and as members of the tribal community.

Tribal people, unlike people living in a caste-based society, had no history of struggles for equality. But they did have a history of struggles for autonomy and independence. This being the case, the constitutional provision granting citizenship rights has been a major breakthrough for scheduled castes and other lower-caste groups. Such has not been the case for the tribes. After all, as observed earlier, citizenship rights, at least in the liberal tradition, have stemmed primarily from the struggle for equality. Tribes were not part of the social system built on the principles of inequality and hierarchy. Hence, they did not have to go through the experience of discrimination—social, political, and cultural—that lower-caste groups, especially the scheduled castes, suffered. Correspondingly, the civil and political rights extended by the Constitution to all citizens were not of great consequence for the tribes, as was the case with the castes ranked lower in the caste hierarchy. The problems of the tribes stemmed not so much from within as from the outside. If at all they were constrained in their enjoyment of freedom and equality, this limitation came more from the outside in the form of an alien state structure, in the form of British rule to begin with and the post-independence Indian state later on. That is, whereas the tribes enjoyed traditional rights in their own society, these rights were curtailed after the incorporation of the tribes into the larger Indian society. Not only were their existing rights curtailed but even as members of the larger Indian society, they had little rights. Rather, they now came to suffer disadvantages similar to those suffered by the groups ranked low in the caste hierarchy. The tribes as collectivities did not aspire or desire to be members of the larger Indian society. That explains why the struggles of tribes have always been for greater control over their land, forests, and other resources through the articulation of a demand for political autonomy rather than a demand for greater participation in and control over state structures and institutions. What they aspired to was space for determining their own way of living, which, of course, also meant exercising ownership and control over resources in the regions where they lived.

The socio-political and economic forces at work in the country have been too strong for the tribes to resist. The result has been the incorporation of the tribes into the larger political and economic system in which they have been given the same rights and status accorded to other citizens, especially after the adoption of the Indian Constitution. It was an achievement only insofar as the larger society made the tribes an integral part and gave them the same rights as those enjoyed by other members. However, the securing of citizenship status hardly mattered to the tribes, for they already enjoyed these rights in their traditional social settings. They already enjoyed the benefits of equal citizenship, including civil, political, and social rights. Of course, it is doubtful if these could be conceived of as citizenship rights since these rights have invariably been local. Citizenship, on the other hand, is by definition national. In other words, the principle of citizenship works in societies based on territory rather than on kinship.

Tribes enjoyed rights not so much in their capacity as members of a political community but in their capacity as members of a society. The political dimension of rights was embedded in the general structure of the social relationships of tribal societies. Political institutions had yet to emerge as distinct and differentiated institutions in tribal societies. Rights in the context of traditional social settings had certain limitations, both in terms of reach and scope. There were also serious limitations in terms of reach and scope on the civil and political dimensions of rights more than on the social and economic dimensions. Certain sections of tribal society, such as women, did not enjoy some of the rights associated with citizenship. Although women enjoyed some of the same rights as men in areas such as the freedom to marry and divorce and the freedom to engage in free movement outside their home and workplace, they also faced many limitations. Customary law and tradition in tribal societies put women at a disadvantage, especially with respect to property and inheritance. To couch it in terms of citizenship rights, women did enjoy the right to personal liberty, the right to freedom of speech and faith, the right to justice, etc. If at all there were limitations in the visibility of these rights in tribal society, it had more to do with the relatively undifferentiated structure of society than rights per se. Women suffered the most serious and glaring disadvantages in the public domain. They did not have a say in the decision-making process affecting the public domain, nor could they hold public office. In other words, women in traditional social settings were denied political rights.

In addition to women, those identified as commoners were also denied such rights. It is interesting that tribal societies, although relatively egalitarian, are not altogether devoid of inequality, especially of rank and status. In many tribal societies, people were divided into chiefs, nobility, commoners, etc. This was the case, for example, among the Lushais of Mizoram, the Kukis of Manipur, the Khasis and the Jaintias of Meghalaya, the Semas and the Konyaks of Nagaland, and the Akas and the Khamptis of Arunachal Pradesh (Misra 1979). These examples are not exhaustive but illustrative. In some tribes, there existed even slaves. The position of commoners in some ways was similar to that of women. There were yet other problems. In many tribal communities, the office of the village chief or head, or even that of the religious head, was confined to members of a particular lineage or clan. Members of other clans and lineages were denied access to these offices, although they were allowed to participate in the decision-making process affecting the village and the community. In this context, citizenship rights conferred by the state have expanded the scope of the rights in tribal societies.

Civil Rights

It is interesting to examine the way and the extent to which tribes have tried to make sense of citizenship rights. This may be discussed with respect to each of the components of citizenship rights. Of civil rights, that is, freedom of thought, expression, faith, occupation, and right to property and justice, the most vulnerable has been the right to property. Many tribal groups in India have been dependent mainly on agriculture for their livelihood. They enjoyed proprietary rights, either individually or communally, over their land. In addition, they also held usufructuary rights over forests and other forms of common property resources. It was in this connection that tribes enjoyed rights of social and economic security, or social and economic rights. In traditional social settings, the right to property was not only an element of the civil domain but also of the social domain. However, the history of the last 200 years, including the post-independence period, has been marked by widespread alienation of land from tribes to non-tribes, especially in regions other than north-eastern India. In the post-independence period, land alienation has been justified on the grounds of the right to property and the right to settlement anywhere in the country. The right to property of one has meant the right to the dispossession of property of another. However, the reality is that this dispossession has invariably been that of tribes. The enjoyment of civil rights on the part of some sections of society has been the root cause of the loss of both civil and social rights by other sections. Of course, tribes have been provided with security in terms of legislation against alienation of land from tribes to non-tribes, which was first introduced during the colonial period. In the post-independence period, characterized by the adoption of policies aimed at bringing about economic development and social justice, such legislation has been enacted in almost all states where tribal groups and communities live. Yet the violation of such rights through legal and illegal means has continued unabated. Tribes, of course, could seek legal redress. But this is a long-drawn-out process and one that requires resources and assistance, which the state did not make available to them Indeed, the right to justice in terms of the restoration of rights was left to the goodwill of a small number of committed bureaucrats and administrators. The right to justice was thus closed off to tribes in the absence of legal and social support from the state. The tribes could not enjoy civil rights as they had little social and economic rights at their disposal.

The other area where tribes have been vulnerable in terms of the exercise and enjoyment of civil rights has been in the domain of freedom of religion. Tribes have their own distinct religions, which in the absence of any specific name have been described as tribal religions. Earlier in the census, tribes were described as animists. They have been drawn to other religions, and have in the course of time become Hindus or Christians or adopted other religions. This process continues even today. However, the conversion of tribes to Christianity has come under severe attack on the ground that conversion invariably takes place under coercion, enticement, fraud, etc. Even if one were to convert of one's own free will, it is said that the act has taken place under the means referred to above, that is, the tribes have supposedly converted because they have been intimidated, coerced, or attacked. A number of states have enacted a freedom of religion law with a view to preventing conversion among tribes. In a sense, then, this is an attack on the civil rights of the tribes as citizens. The question here is not one of conversion alone. Rather, larger issues related to citizenship are involved. It should be remembered that tribes have their own traditional religions. However, they are not being given the right to express and maintain their religious identities. Ever since the adoption of the Indian Constitution, they have either been constrained or coerced into not declaring their traditional religions or into declaring themselves as Hindus. There may be some similarity between the traditional religious practices of the tribes and Hinduism, but that does not make them Hindus. Similarly, there may be some similarity between the traditional religious faith of the tribes and Christianity, but that does not make them Christians. However, Hindutva social and political activists have been exerting pressure, working out strategies, and manipulating situations to ensure that tribes enumerate themselves as Hindus and not as animists or as followers of tribal religions in the census enumeration (Xaxa 2000). Pamphlets were distributed recently in Jharkhand calling for the wiping out the Santhals, the Mundas, and the Oraons for their refusal to enumerate and declare themselves as Hindus.

The other important component of civil rights consistently violated in the case of tribes falls in the realm of language and culture. Article 19(5) of the Indian Constitution grants cultural or linguistic minorities the right to conserve their language and culture. This right is a part of the general right to freedom of speech, expression, and faith. This means that tribes have the right to use their own language, practice their own religion, study their own history, culture, tradition, heritage, etc. Yet tribes have been unable to safeguard and promote their language and culture. The state cannot by law force upon the tribes any other language or culture. Now the state may not have enacted any legislation to this effect, but in actual practice this is what it is doing. There are two main strategies through which this goal is being accomplished. The first strategy is opening schools and imparting instruction through the language and culture of the dominant community. The second strategy is not taking any worthwhile positive steps to meet the constitutional provisions meant for the welfare and benefit of the tribes. It was—and still is—surely asking too much of the tribes to take initiatives for protecting and safeguarding their language and culture on there own. They are economically, socially, and educationally too weak to take any concrete steps in this regard. This explains why there has been so much erosion of tribal languages and cultures in western, northern, and southern India in comparison to the eastern and north-eastern parts of the country. The promotion of language and culture has been left to the tribals. However, because of the lack of control over human, organizational, and financial resources, they have not been able to take effective measures in this direction. It is only when such support has been made available that the tribes have been able to protect and safeguard their cultures and languages (Xaxa 2001).

Political Rights

The exercise of citizenship rights by tribes is the most pronounced in the domain of politics. This is evident in respect of both the right to vote and, more importantly, the right to be elected. However, the moot question is whether tribes would have been in a position to represent themselves as citizens if there were no special provisions for them. The answer is emphatically negative. They undoubtedly would have participated as voters and electors, but certainly not as the elected, capable of exercising political authority. Rather, they would have been represented by non-tribes. Thus, it is not so much the right to be elected as the provision of reservation that has been the most crucial in this context. In the absence of reservation, the citizenship right to be elected would have been meaningless. At least this would have been the case in the early phase of electoral politics, if not today. This can be inferred from the composition of the political parties of all ideological persuasions at the time of independence. In fact, tribes had either no place or a negligible place in the structure of most political parties during this period. There had been little participation of tribes in national political parties, and hence the question of their representation in the absence of reservation would hardly have been possible.

Access to the political decision-making process, which in a democratic set-up is made possible only through election, requires resources (physical and human) that the tribes can-ill afford. Hence, had there been no reservations, there would have hardly been any serious and systematic efforts by political parties to recruit tribes to their cadres. The fact that most political parties were constrained to build cadres from among the tribes was the result of the constitutional provision of reservation for them in Parliament and the state legislatures. Even after cadres have been built from among the tribes, they have, on the whole, been deprived of positions of importance in political parties. Such an attitude or policy cuts across political parties of all persuasions. This is the case not only with political parties but also with organizations like trade unions. The labour force in the tea gardens of West Bengal and Assam consists primarily of tribes. It is another matter that scheduled tribe status has not been accorded to them in Assam. Many trade unions affiliated to different political parties have been working among tea garden workers. Yet the leadership of trade unions has always been in the hands of people from outside the tribes and from outside the labour force. It is only in recent years that one can see a shift in the social composition of the trade union leadership in the tea gardens of Assam and West Bengal. This is a clear pointer to the fact that if there as no obligation of political reservation, the tribes would have been represented largely by non-tribal rather than tribal leaders.

While the provision of political participation helped expand the scope of tribal leadership, and many gained materially and non-materially in the process, such benefits have been limited on scope. After all, in terms of providing space for the protection and welfare of tribes through control over land, forest, and other resources, it has made little contribution. In view of the exercise of political rights from time to time at different levels, there has been, however, heightened political awareness, greater articulation of interests, and increased aspirations among tribal people. Indeed, politics and political mobilization are seen as the only option left for tribes for the protection and promotion of their interests. Hence, it is through political mobilization and political articulation that tribes are attempting to safeguard and promote their social and economic rights today.

Social Rights

It is in respect of social rights that tribes have been the most vulnerable. They enjoyed social security and protection as a result of the control they had over land and other resources. Control over resources was based on either ownership rights or usufructuary rights. These traditional rights are being eroded. The incorporation of tribes into the colonial state structure through annexation also led to their incorporation into the larger system through means of modern systems of transport and communication and the penetration of the forces of capital, market, wage labour, and credit. These processes drastically altered the traditional relationship of the tribes with land, forest, and other resources. Traditionally, these resources were either individually or collectively owned. Even if the tribes had no ownership rights, they had at least usufructuary rights. But the forces unleashed by both the colonial and the postcolonial states led to the steady erosion of the traditional rights of tribes to control and use these resources. This process has taken place in the following ways. The incorporation of tribes into the larger social system opened up tribal areas to people from outside the region as never before. This led to large-scale migration of non-tribal people into tribal areas. This development was not confined to physical movements of people in search of employment, trade, and commerce. It also led to greed, fraud, coercion, and manipulation, and the tribes found themselves increasingly uprooted from their land and being cheated of their resources on an unprecedented scale. The process of land alienation through such means that began during the period of British rule continued unabated in the post-independence period. In order to deal with the problem of land alienation, laws have been enacted in almost all states with tribal populations. In some areas, such acts have been in existence since the British period, such as the Chhotanagpur Tenancy Act, 1908 and the Santhal Parganas Tenancy Act, 1940. In the post-independence period, all states with tribal populations have enacted legislation not only for the prevention of land alienation but also for its restoration. Land protection has also been ensured in certain areas through arrangements such as the Fifth and Sixth Schedules of the Constitution. Despite such acts, tribal lands have continued to pass from tribes to non-tribes. This means that legislative and constitutional provisions alone are not adequate in ensuring the social and economic rights of tribal people. Rather, these provisions need to be complemented and supplemented by effective administrative, infrastructural, and financial support. With of provisions for which certain support systems were made available, for example, the provision of protective discrimination, such arrangements did lead to some results, however inadequate they may have been. But where such measures were non-existent or largely ineffective, legal and constitutional provisions have not led to any desirable results in favour of tribes.

It is not only that effective social and economic rights were not evolved and extended to the tribes, but also that the rights that they already enjoyed, such as those over land and forest, were taken away from them by the colonial state to begin with and by the post-independent Indian state later on. This dispossession has been compounded by official development policies, especially in the post-independence period. Indeed, the major cause of tribal uprooting and displacement has been the state itself. Large-scale industrialization, exploitation of mineral resources, and construction of irrigation dams and power projects have made more tribal people landless, jobless, and homeless than the transfer of land from tribes to non-tribes. During the first three Five Year Plans, more than 50,000 scheduled tribe and 10,000 scheduled caste families were uprooted from the Jharkhand area alone to make land available for the construction of public sector projects. The total number of displaced families would be much higher if one were to take private sector enterprises such as coal mines into consideration (Sengupta 1979). The state projects for national development—whether industry, irrigation, power, or mining—may have benefited the country as a whole and contributed to the general good, but they certainly did not contribute to the good of tribal people. Of course, the state has been sponsoring projects of tribal development but most of these projects are of little consequence if the tribes lack resources.

Conclusion

Tribes in terms of citizenship rights hold the same status as that accorded to members of other communities. Yet in terms of enjoying these rights, the position of tribes in comparison to that of others has been highly vulnerable. As citizens, they have suffered the most mainly because of a legacy of disadvantages, which has been cumulative. The disadvantages have been demographic and historical, on the one hand, and ethnic and structural, on the other hand. Structural disadvantages have economic, political, social, and cultural dimensions, resulting from the subordination of tribes after their incorporation into the larger social system. However, the disadvantages have been far from even in the different components of citizenship rights. In retrospect, one can see some positive developments on the civil and political fronts, especially on the latter front. Yet even these positive developments have not been due solely to civil and political rights per se, but rather due to social rights in the form of reservation and other policy measures meant for the welfare and protection of tribes. At the same time, the provision of such social rights in itself is meaningless if the people who are the intended beneficiaries either lack the required resources or if effective institutional support is not made available to them. Where such support has been made available, tribes have at least made sense of the citizenship rights enshrined in the Constitution. Even here the achievements have, on the whole, been far from satisfactory, especially in regard to civil rights. Results have been most inadequate on the social and economic fronts. Hence, tribes today are increasingly taking recourse to civil and political rights, especially the latter, with a view to promoting, protecting, and safeguarding their social and economic rights and interests.

References

Bakshi, P. M. 2000. *The Constitution of India*. Delhi: Universal Law Publishing Co. Pvt. Ltd. (reprint edition).

Béteille, André. 1974. *Studies in Agrarian Social Structure*. New Delhi: Oxford University Press.

———. 1996. 'The Mismatch between Class and Status.' *British Journal of Sociology* 47 (3): 513–25.

Heater, Derek. 1999. *What is Citizenship?* Cambridge: Polity Press.

Marshall, T. H. 1977. *Class, Citizenship and Social Development*. Chicago: University of Chicago Press.

Misra, B. P. 1979. 'Kirata Karyokiness: Modes of Production in Tribal Societies in North-East India', in *Agrarian Relations in India*, edited by A. Das and V. Nilkant. New Delhi: Manohar, pp. 51–82.

Sengupta, Nirmal. 1979. 'Class Structure in Jharkhand', *National Labour Institute Bulletin* 5: 7–8.

Xaxa, Virginius. 2000. 'Tribes, Conversion and Sangh Parivar.' *Jnanadeepa* 3 (1): 23–35.

———. 2001. 'Tribes and Their Legal Entitlements', in *Human Rights Education in Schools and Colleges*, edited by Herman Castelino. Anand: Gujarat Sahitya Prakash.

Chapter 5

Tribal Movements: Rethinking in a Comparative Perspective

Tribal movements have been one of the most extensively studied developments in tribal studies. What accounts for this focus is the pervasiveness of tribal people's resistance against alien forces ever since they came under British rule and administration. Raghavaiah has made a chronological list of 70 'tribal revolts' from 1778 to 1971 (Raghavaiah 1979). A survey of tribal movements conducted by the Anthropological Survey of India has identified 46 tribal movements in India during 1976–77 (Singh 1993). As early as 1945, *Man in India*, an anthropological journal, brought out a special number on tribal rebellions. This chapter examines the salient features of studies of tribal movements and critiques some prevailing ideas on tribal movements in sociological and social anthropological literature.

Most studies of tribal movements are in the nature of ethnographic accounts. Hence, studies have tended to be mostly descriptive rather than analytical. These studies have led to certain kinds of broad generalizations. For example, tribal movements have been, by and large, associated with tribes that are numerically large and relatively advanced. The relatively advanced tribes are associated with tribal groups that make their living from agriculture, especially settled agriculture. Conversely, it is held that small, isolated, and primitive tribes have little potential for launching and sustaining a movement. At best, they are expected to react to issues over land and forests on which they are greatly dependent for their livelihood (Singh 1982). Tribal movements have also been characterized as spontaneous events, thereby lacking organizational structure. Hence, attempts have been made to show the existence of organization in such movements (Guha 1983). Others have pointed to tribal social and political structures as the key elements of the organizational structure of tribal movements (Shah 1990).

One of the concerns in the study of tribal movements in India has been to provide a system of classification. To construct a typology of such diverse social movements is a difficult task. Generally, the movements have been classified keeping in mind geographical and/or chronological factors. Hence, they have been classified as region-specific, such as those of north-eastern India, eastern India, central India, etc., or on the basis of chronology, as those of the pre-independence and the post-independence periods. Movements have also been classified on the basis of their objectives and aims as revolts, rebellions, reform movements, insurrections, etc. On the basis of ideology, movements have been classified either as conservative and reactionary or as progressive and revolutionary (Shah 1990; Fuchs 1965; Mahapatra 1972). The most common way of classifying movements has been on the basis of issues, such as forest, land, ethnicity, culture, identity, political autonomy, social mobility, etc. There is a broad agreement among scholars over employing such a system of classification for tribal movements. However, such a system of classification in tribal contexts is far from neat, and often overlaps with other issues. The classifications thus reflect more of the dominant themes and/or issues underlying the movements. With regard to movements broadly described as pertaining to the cultural domain, there are various interpretations. Whereas Singh (1983, 1993) uses the terms 'Sanskritization' and 'cultural' to describe such movements depending on the overall thrust, Sinha (1972) employs the term 'reform' in place of 'Sanskritization' and 'ethnicity' in place of 'cultural' to refer such movements.

Early Tribal Struggles

Tribal struggles of one type or another continued all through the colonial period. However, it has been difficult to fit the early tribal struggles into any of the types mentioned above. Tribal struggles of the colonial period have come to be described more on the basis of their character than on the basis of the issues on which they were based. Indeed, the issues were too fuzzy to be clearly conceptualized. Thus, struggles dating from this period have been commonly described as revolts, rebellions, or insurrections. At times, they have been described in terms of messianic or millenarian movements. Such descriptions tend to emphasize the fact that the early tribal struggles had more to do with the issue of colonial rule and administration. Hence, it would not be incorrect to describe these early struggles as struggles for autonomy of a certain kind. Of course, the demand for autonomy articulated in the post-independence period has been different in a very substantive sense from the issues that provoked the struggles of the British period. Yet there was an overlapping, at least in the spirit if not in the letter, as may be inferred from the way in which the early struggles were described by the colonial administration. The terms commonly used to describe these struggles were revolt, resistance, and insurrection. Even scholarly historical works adhere to such descriptions. An interesting aspect of the tribal struggles of the period is that in the postcolonial official history of India, these have come to be seen as a part of the freedom struggle, thus equating the struggles of the tribes with the struggles of the non-tribes against British rule. In the process, what is often forgotten is that the tribal struggles were not only against the British but also against non-tribal people, who in turn were engaged in a struggle against the British. The non-tribals were invariably seen as people who took advantage of the administrative and other structures created by the British and who evicted tribes from their lands and deprived them of their resources through fraud, coercion, and debt bondage. In fact, tribal struggles were fought against non-tribal people far more intensely, as they were seen to be the immediate exploiters and oppressors of tribal people.

Sanskritization Movement

Colonial rule and administration became deeply entrenched in tribal areas after the failure of the early struggles to replace existing arrangements. This led to an unprecedented expansion of roads, railways, and other means of communication, which further opened up the space for an influx of people from outside into the tribal areas. The alienation of land from tribes to non-tribes continued unabated. However, the initial shock that they had suffered had been absorbed by now. There was now greater communication and interaction between tribes and non-tribes. The non-tribals were now seen as people who occupied positions of power and influence, and hence were superior to the tribals. The non-tribals looked down on the tribals and criticized the latter's way of life and values. The result was that the tribals began to emulate the way of life and values of the non-tribals. This process has been referred to as Sanskritization in social anthropological literature. In due course, the urge to emulate became so strong that it took the form of movements that have been described as Sanskritization or social mobility movements. The origins of these movements can be traced to the beginning of the twentieth century; they gained momentum in the 1920s and 1930s. The movements went by different names in different regions and among different tribes. Thus, in Gujarat, the movement was popularly known as the Devi movement (Hardiman 1987). Among the Oraons of Chotanagpur, it was called the Tana Bhagat movement (Ekka 1972, 2003; Sachchidananda 1978). Among the Hos and the Santhals, it was known as the Haribaba and the Sapha Hor movements respectively (Ekka 2003; Singh 1992; Areeparampil 1993). Yet these movements were strikingly similar in many ways, and hence are commonly known as the Bhagat movements. The Bhagat movements have also been widespread among many tribes in central and western India (Bhatt 1983; Danda 1983; Mann 1983; Lal 1983).

Those who regard such movements as evidence of Sanskritization view these primarily as evidence of attempts to increase social mobility. Indeed, the two terms have been used almost interchangeably. Many scholars have taken this line. To my mind, however, the movements are is imbued not so much by the desire for social mobility as by the desire to gain respectability in the eyes of the larger society. This is, of course, not to deny the linkage between Sanskritization and social mobility. However, where this has been so, it has been more of a political than a cultural movement. That is, Sanskritization as a cultural process may have been at work among the tribes but it is not linked to issues of social mobility. In fact, the open articulation of social mobility has been made only in areas where the Sanskritization process has already taken place widely. Among members of such groups, there has been a widespread movement for the recognition of a higher status, such as Kshatriya status. Since such articulations have often been in the form of demands for their enumeration as Kshatriya in the census or official records, the movement assumes a political rather than a cultural form. The cases of the Kudmi Mahtos of Jharkhand and the Koch Rajbongshis of Assam and northern Bengal are illustrative of this phenomenon (Singh 1983; Basu 2003).

There is no doubt that such movements have been active vehicles for Sanskritization. However, to describe them merely as that would be an over- simplification. Rather, these movements are more in the nature of a critique of their society. After all, tribes have been steadily losing control over their land and other resources, and they have had to face extreme exploitation. Although they have resisted these conditions from time to time through armed revolt, they have been unsuccessful in changing their current circumstances. They have failed to redress their grievances through legal action not only because the courts are located far away from the places where they live, but also because the entire legal process, including the attitude of the functionaries of the court, is hostile to them. Hence, tribals have sought a solution to their problems by directing attention towards their own societies. That is, they have sought a solution not in terms of changing the relations between tribes and non-tribes but by initiating change within their own societies. The movements have been directed primarily at reforming the inner workings of their own societies, including their religions since religion is intimately related with their social structure (Ekka 1972, 2003; Sachchidananda 1978; Hardiman 1987).

Tribal religion has many spirits, both benevolent and malevolent, which are invariably treated as the custodians of the peoples' well-being. However, neither benevolent nor malevolent spirits, despite the prayers and sacrifices offered to them, had succeeded in protecting the tribes from the suffering and misery that had befallen them since the inception of British rule. That explains why the reform or revitalization movements came to focus on religion. This proceeded in two directions. The first was an increasing trend of replacing tribal gods and spirits with the gods and deities of more powerful religions, especially Hinduism and Christianity. The second was increasing pressure to return to the original form of tribal religion. This meant reverting to the one supreme god whom they worshipped in their traditional religion (Ekka 1972, 2003; Mahapatra 1972). The principal aim of these reform or revitalization movements was to purge tribal society of those elements that provoked the scorn and disdain of the larger society. They were also imbued by prophetic visions of ushering in a new social order. Although purificatory in nature, they were also expressions of protest and rebellion against the colonial rulers and their allies, such as zamindars. The movements encouraged non-payment of revenue, rent, and tax, and a withdrawal from forced labour. This is seen most clearly in the Tana Bhagat movement among the Oraons, the Devi movement among the tribes of south Gujarat, and the Haribaba movement among the Hos (Ekka 2003; Hardiman 1987; Areeparampil 1993; Singh 1992).

Again, there is no denying the fact that some movements have been overwhelmingly driven by the ethos of Sanskritization. The same could not be said, however, of such movements as a whole. Indeed, many such movements, though apparently characterized by features resembling those of the Sanskritization process, were, in fact, driven more by a desire to reform tribal society so as to be able to withstand the forces of change rather than through Sanskritization or emulation, as argued by many scholars. That is, there were other discernible strands in such movements—reform and ethnicity being the two major ones (Ekka 1972, 2003). Indeed, the three stands—Sanskritization, reform, and ethnicity—overlap a great deal with each other. Hence, the description of such movements as being driven only by Sanskritization glosses over other important dimensions.

Cultural and Identity Movements

There is some confusion about the cultural dimension of tribal movements, with the movements being linked to the Sanskritization process, namely the Bhagat movement. However, in a very substantive sense, the Bhagat movement is antithetical to the Sanskritization process, for the aim of the former is to revitalize tribal society, which has experienced many kinds of changes resulting from contact with the outside world. The aim of the Bhagat movement is to restore the original status and distinctive features of tribal societies lost because of the influence of Sanskritization. Hence, it is a reform movement aimed at restoring tribal society to its pristine state, that is, an attempt to go back to the original tribal religions, social customs, practices, and values. For example, some tribes like the Santhals, the Hos, and the Oraons attempted to return to pre-Hindu and pre-Bhagat forms of tribal religion. These attempts are known by different names, such as the Sarna Dharam, the Sari Dharam, and the Adi Samaj movements. These are described as revitalization movements in the anthropological literature. Examples are the Adi Samaj movement initiated by Lakhi Bodra, Dupud Dharam, and others among the Hos and the movement initiated by Raghunath Murmu among the Santhals (Areeparampil 1993; Mahapatra 1983). Revitalization movements are also found among some tribal groups in the north-eastern region. The most notable among these are the Seng Khasi movement among the Khasis and the Zeliangrong movement among the Zemeis, the Liangmeis, and the Rongmeis Nagas (Kabui 1982; Natarajan 1982).

In such movements, the question of identity is brought to the forefront. The issue of identity, which had been dormant in earlier tribal movements, has added a new dimension to tribal movements. Identity issues have been articulated in both cultural and political terms.

Indeed, political issues have been instrumental in the rise of autonomy movements among different tribal groups in India. However, even where they enjoy autonomy in the form of a separate state, the issue of identity articulation continues to be important. In tribal states or regions where the non-tribal population is large and where the steady inflow of non-tribal population continues, the issue of tribal identity has been sharply articulated. This is the case in north-eastern India (Kumar 2005). In the tribal state of Meghalaya, there was recurrent tension in the recent past between tribals and non-tribals over the issue of identity. The fear of being reduced to a minority in their own land was intensely articulated. In short, the articulation of identity has been taking place on two planes—political and cultural. The former is an external articulation of identity since it hinges primarily on tribal–non-tribal relations. The latter is an internal articulation of identity. Nevertheless, the two kinds of articulations are linked with each other. In the north-east, the identity issue is more of a political than a cultural articulation as the cultural distinctiveness of tribal groups has never been at stake. When it has been at stake, it has provoked a demand for a separate hill state. In contrast, the cultural distinctiveness of tribal groups has been eroded as the result of increased contact with the larger Indian society. The erosion became intense and widespread with the consolidation of British rule. Hence, the demand to return to their cultural roots has been articulated with the aim of asserting their distinct and separate identity. The means through which this goal was achieved took a predominantly cultural rather than a political turn.

Land- and Forest-based Movements

Land- and forest-based movements historically have been more widespread in eastern, central, and western India. In southern India, such movements have been a recent phenomenon. The land and forest movement waged by the Adivasi Gotra Samaj in Kerala under the leadership of Janu has received wide attention. In north-eastern India, such movements, excepting those in the plains areas have been rare. This is mainly because of the control that the tribals still enjoy over land and forests, especially in the hill region. The same has not been the case in the plains.

Discontent over land and forest issues among the tribals began with the coming of British rule. The British introduced new policies and administrative arrangements in the tribal areas. Side by side, the region was also opened to market forces. All these changes led to loss of control over land by the tribals. At the same time, restrictions were imposed on the use of forest resources, which caused widespread resentment against British rule and the aliens who had appropriated the resources of the tribals. The tribal revolts or rebellions that occurred throughout the nineteenth and the early twentieth centuries were fuelled mainly by these grievances. These grievances, however, did not acquire the form of a land- or forest-based movement as the issues were closely connected with the overall social destabilization brought about by colonial rule and alien administration. Since those who had usurped tribal land were outsiders (*dikus*), including British administrators, the movement took the form of a revolt against British rule and the outsiders, especially landlords, moneylenders, and traders. Prominent among these movements were the Kol insurrection, the Bhumij revolt, the Santhal rebellion, the Kherwar movement, and the Birsa movement. After these initial outbursts, there was a lull in the articulation of discontent. Even if popular outbursts did take place, these got tagged, and even became secondary to the reform movement witnessed by tribal society during this period. Such movements have been referred to as the Bhagat movements in the anthropological literature, as noted earlier. However, as the Indian national movement gained momentum, agrarian and forest issues came to the forefront among the tribes, but this time with a difference. The movements now were part of the larger movements of civil disobedience and non-cooperation launched by the Congress and other nationalist leaders. Agrarian movements during the larger national struggle not only took the form of the non-cooperation and civil disobedience movements against the colonial state but also struggles against exploitation and oppression by landlords. The latter, both in the colonial and postcolonial periods, were mobilized primarily by left-wing political parties. The broad aims of such struggles were security of tenure, reduction of rent, increased share of produce, etc. In these struggles, tribals aligned with other exploited classes of agrarian society (Sen 1972, 1982; Singh 1983; Dhanagare 1983). Agrarian-based movements in post-independence India also assumed the form of forcible harvesting of crops (on lands that once belonged to tribals, but which had been taken away from them through fraudulent means), repudiation of debts, and implementation of the prescribed minimum wage. Such struggles were directed against moneylenders and landlords, and were widespread in parts of eastern and western India in the 1960s and 1970s (Iyer and Maharaj 1986; Adhikary and Bhattacharya 1983; Kulkarni 1983).

The large majority of tribes in India are dependent on agriculture for their livelihood. Hence, it is generally argued that those practising settled agriculture are no different from the neighbouring non-tribal peasants. Hence, many scholars treat tribal movements as peasant movements. Peasant leaders too have described them as aboriginal kisans. Such an approach, however, tends to gloss over the diversities of tribal social formations of which tribal movements have been a part. The patterns of leadership and political mobilization among the tribes differ from those found among peasants. The difference is attributed to the characteristic features of tribal society, such as physical or geographical concentration in certain areas, social and political organization, and relative isolation from the so-called mainstream. Tribal movements are also seen as being different from peasant movements in the sense that tribal movements are both land and forest based. Tribal movements also exhibit a greater sense of community consciousness (Singh 1983).

In studies of land- and forest-based movements, tribes have invariably been treated mostly as peasants and forest dwellers. Their concerns have been addressed exclusively in terms of securing their economic livelihood. Where such movements did take place, they were invariably led by outside agencies, such as political parties, activists, and social workers. Where tribals on their own have taken up such issues, they have invariably been part of larger issues, such as demands for political autonomy or for greater control over land, forests, and other resources. This has been the case with movements in both pre-independence and post-independence India.

As observed earlier, one of the features of agrarian movements among tribes has been resistance to the alienation of land from tribes to non-tribes. In recent years, the resistance to land alienation has increased in momentum, not so much in respect of alienation of land from tribes to non-tribes but in respect of alienation resulting from large development projects, sponsored both by the state and by private capital. The effect or outcome of land alienation resulting from such projects is referred to as displacement. There has been massive displacement of tribal people throughout the post-independence period because of the exigencies of national development. However, the benefits of these development projects have not accrued to tribal people. Rather, they have been further impoverished because of their displacement. Hence, resistance against such projects has increased in the tribal regions in recent years. Many such movements are found in different parts of the country today. Some of these movements like the Koel Karo struggle in Jharkhand have been in existence for more than three decades. Some other important ongoing movements are the struggles of tribal people in Kashipur, Lanjigarh and Kalinga (Orissa), Nagarnar, and Mehendikheda (Chhattisgarh). The key players in these struggles have been the tribals themselves. NGOs and activists have also begun to support these struggles. Another noteworthy struggle against displacement (eviction of tribals from forests) that has received national attention in recent years is the struggle of tribals in Kerala, which is being spearheaded by a woman leader called Janu, under the auspices of the Adivasi Gotra Samaj.

Of the movements against displacement, the one that has received worldwide attention is the Narmada Bachao Andolan. This movement has attracted wide participation from wide a cross-section of people. The key players besides the tribals are NGO and middle-class activists from different occupational and professional backgrounds. Indeed, the role of the latter has been so dominant that the movement has come to be strongly associated with these middle-class activists and NGOs. One of its singular achievements is that it has made a serious critique of the developmental model of the state, which continues to be in vogue even today. It has opened up an alternative vision of values and development.

Autonomy Movements

Of all the tribal movements, movements for autonomy have drawn wide attention. Autonomy movements are an expression of the larger articulation of identity by tribals. Such movements have been widespread in the north-eastern region, where they have been based on intense mobilization. Elsewhere in the country, they have been confined to certain pockets of eastern, central, and western India. In Gujarat, tribes living in the southern part of the state demanded autonomy. In Madhya Pradesh, the Gonds demanded a separate state of Gondwana in the 1950s. Yet in Gujarat and Madhya Pradesh, the demands failed to become organized movements (Singh 1983; Desai 1983). The only autonomy movement worth its name in mainland India has been the movement for the separate state of Jharkhand. This has been one of the longest fought struggles and has passed through many historical ups and downs. Since its inception in the 1930s, it has undergone considerable changes in its character, organization, strategy, and mobilization. Since the 1970s, the earlier ethnic character of the movement changed into a regional character. Autonomy movements among tribes in mainland India have thus been few. With the exception of Jharkhand, the demand for autonomy in the form of a separate state here has been primarily employed as a strategy to gain some benefits or concessions for tribal people (Desai 1983).

In the north-eastern region, there have been many autonomy movements in various tribal areas and among different groups. Some of the prominent ones were waged by the Nagas, the Mizos, the Khasis–Jaintias–Garos, the Boros, and other tribal groups in Tripura. Some movements such as those of the Nagas, the Mizos, the Bodos, and the Tripuri tribes have gone to the extent of demanding secession from India. Although some degree of autonomy has been granted to tribes and/or regions making such a demand, the movement for autonomy has not come to a stop. Rather, it has spread to new regions and tribes. Further, where the demand for autonomy has been dormant, or where it has not yet been granted, or where autonomy has fallen short of the aspirations of the people, there has been either a renewed interest in the movement (Gondwana) or the movement has become more intense and aggressive. The latter is the case, for example, with the Bodos and the Karbis in Assam and with the Tripuri tribals in Tripura.

The beginnings of the tribal autonomy movement in the form of a demand for a separate state can be traced to the period preceding the end of British rule. However, in nascent form, it had already appeared in the 1920s. The tribal leadership in Jharkhand and Nagaland had raised the issue during the visit of the Simon Commission to India. The difference between the two was that while the tribal leadership in Jharkhand articulated the demand for a separate state within India, Naga leadership demanded to a sovereign state. After India became independent, the movements for separate states in both Jharkhand and Nagaland grew more intense, involving mass mobilization and participation. The movement in Nagaland turned violent, and the Government of India had to deploy the army. Simultaneously, as a part of the strategy to contain the movement, the Nagas in 1963 were granted separate statehood within the Indian Union. The Nagas were, however, unhappy with the limited autonomy granted them, and the struggle for sovereign autonomy continues until this day. In Jharkhand, the demand was not given serious consideration, and the struggle, as in the case of Nagaland, continues.

Barring Nagaland, organized movements for autonomy either within the Indian Union or outside it, were, on the whole, absent in the north-eastern region in the early years after independence. By the 1960s, however, the movement for a separate state began to gain momentum among the tribes living in the hilly regions of Assam. The regions comprised the Khasi, Jaintia, and Garo Hills (which form the present-day state of Meghalaya), the Lushai Hills (which are part of the union territory of Mizoram), and the hill districts of the North Cachar and the Mikir Hills (in the present-day state of Assam). Paradoxically enough, the regions and/or districts already enjoyed a certain degree of autonomy, which was not enjoyed by the tribes, including the hill tribes in the other states of the region, such as Tripura, Manipur, and Arunachal Pradesh, namely the erstwhile NEFA. Indeed, the earlier demand in Tripura was for an autonomous regional council under the Sixth Schedule of the Constitution.

Despite the autonomy enjoyed by the tribes of these areas, the demand for separate statehood continued. Soon the demand was transformed into a strong movement, based on intense mobilization and widespread participation. In Mizoram or the erstwhile Lushai Hills district, the movement moved in the direction of secession from India. The demand for a separate state was eventually conceded, and the state of Meghalaya and the union territory of Mizoram were created. In Mizoram, even after the granting of greater autonomy, the struggle for secession from India continued. Indeed, the movement during this period came to be marked by a widespread armed struggle. The struggle came to an end with the signing of an accord in 1986.

How do we explain the autonomy movements that took place during this period? After all, as noted earlier, the tribes already enjoyed some degree of autonomy. The answer lies in the fact that the autonomy granted to the tribes had in practice many limitations. To begin with, though autonomous district councils gave some powers to the tribes with respect to determining many aspects of their life, the powers granted to the institution suffered from certain limits, which came from two sources. The first source was the limited power given to the district council. The laws enacted by the council had no effect until the governor gave his assent. It could be dissolved as and when it was inconvenient to the state government. Even its acts and regulations could be annulled or suspended. The district council enjoyed certain executive functions in respect to several areas listed in the Sixth Schedule, but without any mandatory role of engagement with development activities. The sources of revenue for the district council and the regional council had been specified in the Constitution, but some of the major sources of revenue were under the control of the state government (Chaube 1973). The second—and more important—source of the limit to the powers of the district council came from the actual politics of the state of which the tribal regions formed a part. One of the major aspects of state-level politics was the deep divide between tribes and non-tribes, or plains people and hills people. As noted earlier, state-level politics was dominated by the non-tribes, which were not only numerically dominant but also socially and economically powerful. Not only did they dominate the tribes, they were also able to thwart and scuttle the functioning of the district council by blocking grants and creating other impediments (Chaube 1973). The district council thus failed to meet the needs and aspirations of the tribal people. The demand and movement for a separate state emerged out of these circumstances.

An interesting question is why the autonomy demands in the regions outside of north-eastern India (with the exception of Jharkhand) could not be translated into organized and sustained movements. In other words, what made the autonomy movement in Jharkhand sustainable?

One can identify a number of factors that were present in the case of Jharkhand and absent elsewhere in the region. One was that the tribes in Jharkhand had a long history of struggle against colonial rule and outside exploiters. This helped them to forge a shared identity despite various differences. They continued to suffer exploitation and domination in the post-independence period, and this served to further reinforce their sense of shared identity. In sharpening this identity, the role of a tiny but articulate middle class was crucial. It was the tribal middle class that experienced discrimination and domination most intensely in terms of access of the supposed benefits of development, employment, trade, commerce, etc. This helped to keep the struggle for autonomy alive. Further, increased communication among the middle classes in different places was facilitated by the growth of towns in the tribal region. Equally important was the strong sense of identification with their own language and culture, which sharpened the sense of difference between tribes and non-tribes. Indeed, wherever there has been a loss of language and culture, the movement for autonomy has generally been weak. This sense of common identity was further strengthened by identifying with a distinct religion, especially Christianity. Indeed, Christianity contributed to the vitality and vibrancy of tribal languages and cultures in many ways. The movement for autonomy in Jharkhand was thus rooted in cultural differences and uneven development between tribes and non-tribes. The factors that we see in Jharkhand were absent in other parts of mainland India. That explains why autonomy movements in other parts of mainland India could not sustain themselves. The mark of identification of tribes such as the adivasis was undoubtedly an important and necessary factor in the articulation of a distinct identity, but this was not sufficient for transforming it into a process of actual political mobilization.

It is interesting to note that where tribes have experienced increasing marginalization as evidenced in the loss of control over land, forests, and other resources, and where there has been no provision for autonomy whatsoever, autonomy movements have generally been weak. The tribal situations in central, western, and southern India bear witness to this fact. The only exception has been Jharkhand, in eastern India. This means that pauperization and expropriation do not necessarily lead to the emergence of an autonomy movement. These factors are important but not sufficient causes, as one can see from cases other than that of Jharkhand.

Unlike the situation in central, western, and southern India, movements for autonomy have been widespread in the north-eastern region. The issue of autonomy has also been fought over most fiercely in this region. Interestingly, the demand for autonomy in the north-east has its genesis not so much in the marginalization of tribes as in the fear of losing their distinct language, culture, customs, and traditions on account of the domination of an alien language and culture (Chaube 1973; Das 1982; Goswami and Mukherjee 1982). This has been the case even where there has been large-scale marginalization of tribes, as seen with the Bodos and the Tripura tribals. The tribes in Tripura, which constituted a majority until independence, became a minority after independence because of the exodus of Bengalis from former East Pakistan, now Bangladesh. This unprecedented demographic shift also led to the erosion of control over land, forests, and other resources. In the Bodo-inhabited areas of Assam too, the migration of the Assamese and the Bengalis has been going on since the period of British rule. As a result, large-scale alienation of land from tribes to non-tribes has continued for nearly a century. This has not only affected the demographic but also the cultural patterns of the region. The Assamese and Bengali languages were adopted and assimilation into Hinduism and its social structure took place, as has been the case with the Jharkhandi tribals. Yet that the cultural dimension, especially the issues of language and script, remained at the heart of the movement (Mukherjee and Mukherjee 1982; Mukherjee and Singh 1982).

In short, the demand for autonomy in the north-east was based on the issue of identity. The social origins of autonomy movements and the degree of their mobilization thus lay not so much in pauperization and marginalization as in other factors. In this case, Sinha's formulation is pertinent in this context. In his view, tribal solidarity is positively correlated with factors such as ecological and socio-cultural isolation vis-à-vis the core peasantry, a certain numerical strength to provide striking power, location near an international border, a certain level of literacy and education to provide elite leadership, historical experience of conflict, opportunity for the 'political rank path', and limited scope for economic rewards (Sinha 1972). At the root of such movements lies a strong sense of identity politics. Autonomy movements by ethnic minorities have thus not been all of the same nature, level, and character. Broadly, they have been articulated in the form of demands for separate states within the Indian Union. It is when these demands have not been met that the demand has been raised for secession from India in the form of establishing a sovereign state. This is not to deny that some movements have been secessionist in character from the outset. However, such movements have been few.

To conclude, tribal movements—understood in the sense of movements articulated and engaged in by tribal people—have been, on the whole, heterogeneous and amorphous in nature, varying in character and orientation. This is because they have invariably tried to address a wide range of issues. This has been particularly true of movements organized and mobilized by the tribal people themselves. In contrast, movements initiated by forces outside tribal society, that is, political parties and activists, have been concerned primarily with livelihood issues. Hence, the construction of typologies has been one of the major concerns in the study of tribal movements. A typology worked out on the basis of issues has been the most widely accepted one. However, the study of tribal movements primarily in terms of issues fails to capture the other intricate dimensions of such movements. This is the case with all tribal movements, irrespective of whether they are described as being concerned with autonomy, identity, cultural or Sanskritization issues, or control over land and forest resources. This is because of the diffused character of tribal society in which different aspects are intricately interwoven, making it difficult to determine where one begins and the other ends. These intricacies are often lost sight of when tribal movements are studied solely in terms of issues.

References

Adhikary, A. K. and R. K. Bhattacharya. 1983. 'The Extremist Movement: An Appraisal of the Naxalite Movement with Special Reference to Its Repercussion Among Tribes', in *Tribal Movements in India*, Vol. 2, edited by K. S. Singh. New Delhi: Manohar, pp. 119–127.

Areeparampil, Mathew. 1993. 'Socio-cultural and Religious Movements Among the Ho Tribals of Singhbhum District of Bihar', in *Continuity and Change in Tribal Society*, edited by M. Miri. Shimla: Indian Institute of Advanced Study, pp. 396–436.

Basu, Swaraj. 2003. *Dynamics of a Caste Movement: The Rajbansis of North Bengal 1910–47*. New Delhi: Manohar.

Bhatt, Vikas. 1983. 'A Baba in Bastar', in *Tribal Movements in India,* Vol. 2, edited by K. S. Singh. New Delhi: Manohar, pp. 223–41.

Chaube, S. K. 1973. *Hill Politics in North-East India.* New Delhi: Orient Longman.

Danda, A. K. 1983. 'Gahira Guru and His Sant Samaj Movement', in *Tribal Movements in India,* Vol. 2, edited by K. S. Singh. New Delhi: Manohar, pp. 197–208.

Das, N. K. 1982. 'The Naga Movement.' in *Tribal Movements in India,* Vol. 1 edited by K. S. Singh. New Delhi: Manohar, pp. 39–52.

Dasgupta, Pranab Kumar. 1983. 'The Adi Samaj Movement Among the Ho', in *Tribal Movements in India,* Vol. 2, edited by K. S. Singh. New Delhi: Manohar, pp. 93–107.

Desai, I. P. 1983. 'The Tribal Autonomy Movement in South Gujarat', in *Tribal Movements in India,* Vol. 2, edited by K. S. Singh. New Delhi: Manohar, pp. 243–59.

Dhanagare, D. N. 1983. *Peasant Movements in India 1920–50.* New Delhi: Oxford University Press.

Ekka, P. 1972. 'Revivalist Movements Among the Tribals of Chotanagpur', in *Tribal Situation in India,* edited by K. S. Singh. Shimla: Indian Institute of Advanced Study, pp. 424–31.

———. 2003. *Tribal Movements: A Study in Change.* Pathalgaon: Asha Deep.

Fuchs, S. 1965. *Rebellious Prophets: A Study of Messianic Movements in Indian Religions.* Bombay: Asia Publishing House.

Goswami, B. B. and D. P. Mukherjee. 1982. 'The Mizo Political Movement', in *Tribal Movements in India,* Vol. 1, edited by K. S. Singh. New Delhi: Manohar.

Guha, R. 1983. *Elementary Aspects of Peasant Insurgency in Colonial India.* New Delhi: Oxford University Press.

Hardiman, D. 1987. *The Coming of the Devi.* New Delhi: Oxford University Press.

Iyer, K. Gopal, and R. N. Maharaj 1986. 'Agrarian Movements in Tribal Bihar (Dhanbad), 1972–80', in *Agrarian Struggles in India After Independence,* edited by A.R. Desai. New Delhi: Oxford University Press, pp. 330–61.

Kabui, G. 1982. 'The Zeliangrong Movement: An Historical Study', in *Tribal Movements in India,* Vol. 2, edited by K. S. Singh. New Delhi: Manohar, pp. 53–66.

Kulkarni, D. S. 1983. 'The Bhil Movement in Dhulia District', in *Tribal Movements in India,* Vol. 2, edited by K. S. Singh. New Delhi: Manohar, pp. 261–71.

Kumar, Nikhlesh. 2005. 'Identity Politics in the Hill Tribal Communities in North-Eastern India'. *Sociological Bulletin* 54 (2): 195–271.

Lal, R. B. 1983. 'Socio-Religious Movements Among the Tribals of South Gujarat', in *Tribal Movements in India,* Vol. 2, edited by K. S. Singh. New Delhi: Manohar, pp. 285–308.

Mann, R. S. 1983. 'Structure and Role Dynamics Among the Bhils of Rajasthan: A Case of the Bhagats', in *Tribal Movements in India,* Vol. 2, edited by K. S. Singh. New Delhi: Manohar, pp. 309–24.

Mahapatra, L. K. 1972. 'Social Movements Among Tribes in India', in *Tribal Movements in India,* Vol. 2, edited by K. S. Singh. New Delhi: Manohar, pp. 399–409.

Mahapatra, Sitakant. 1983. 'Raghunath Murmu's Movement for Santal Solidarity', in *Tribal Movements in India,* Vol. 2, edited by K. S. Singh. New Delhi: Manohar, pp. 129–59.

Mukherjee, Bhabanand and K. S. Singh. 1982. 'Tribal Movements in Tripura', in *Tribal Movements in India,* Vol. 2, edited by K. S. Singh. New Delhi: Manohar, pp. 317–38.

Mukherjee, D. P. and S. K. Mukherjee. 1982. 'Contemporary Cultural and Political Movements Among the Bodos of Assam', in *Tribal Movements in India,* Vol. 2, edited by K. S. Singh. New Delhi: Manohar, pp. 253–80.

Natarajan, N. 1982. 'Cultural Revivalism Among the Christian Khasis of Meghalaya', in *Tribal Movements in India,* Vol. 2, edited by K. S. Singh. New Delhi: Manohar, pp. 151–72.

Raghavaiah, V. 1979. 'Background of Tribal Struggles in India' in *Peasant Struggles in India,* edited by A. R. Desai. New Delhi: Oxford University Press, pp. 12–26.

Sachchidananda. 1978. 'The Bhagat Movement in Chotanagpur', in *Indian Movements: Some Aspects of Dissent, Protest and Reform,* edited by S. C. Malik. Shimla: Indian Institute of Advanced Study, pp. 159–90.

Shah, Ghanshyam. 1990. *Social Movements in India.* New Delhi: Sage Publications.

Sen, Sunil. 1972. *Agrarian Struggle in Bengal 1946–47*. New Delhi: People's Publishing House.

———. 1982. *Peasant Movements in India: Mid-Nineteenth and Twentieth Centuries*. Calcutta: K. P. Bagchi.

Singh, K. S. (ed.). 1982. *Tribal Movements in India*. Vol. 1. New Delhi: Manohar.

———. (ed.). 1983. *Tribal Movements in India*. Vol. 2. New Delhi: Manohar.

———. 1992. 'Haribaba and His Movement in Chotanagpur', in *Tribal Transformation in India*, Vol. 3, edited by B. Chaudhuri. New Delhi: Inter-India Publications, pp. 344–56.

———. 1993. 'The Problem in Marginalized Tribals.' *Seminar* 412: 1–7.

Sinha, Surajit. 1972. 'Tribal Solidarity Movements in India: A Review', in *Tribal Movements in India*, Vol. 2, edited by K. S. Singh. New Delhi: Manohar, pp. 410–23.

Chapter 6

Empowerment: Forms, Limitations, and Assertions

The idea of political empowerment is quite new. As a reality, however, it is as old as the system of democracy or populism. Democracy is, after all, regarded as the rule of the people, by the people, and for the people. Similarly, populism aims to give more space and rights to the people. Empowerment is thus linked with equality, liberty, and fraternity. Hence, even liberation struggles against colonial rule may be seen as a step in the direction of empowerment. In the colonial context, the demand for empowerment may be traced to the demand for greater participation by natives in government and later in the struggle for liberation. It is, of course, a different matter that subsequent gains resulting from this demand were monopolized by native elites, as was the case in India. No doubt, the powers granted to natives during the colonial period constituted a process of empowerment, but this was of a limited nature. Power was, after all, not within the reach of subaltern people. Post-independence India broadened the scope of power-sharing for the people. Yet people's participation has fallen far short of popular expectations and aspirations. It has remained confined to the level of formality. As far as actual control is concerned, ordinary people have very little say and a limited role to play. This situation has led to a demand for an increased share in power, both economic and political. It is hardly surprising that increasing demands for political power are being made, especially by the disadvantaged, namely workers, peasants, women, youth, dalits, and tribes. Since it is not possible to address here the issue of empowerment of such diverse categories of people, an attempt has been made in this chapter to discuss the issue of the empowerment of tribes in India.

In sociological literature, the notion of power is used broadly in two senses. One is in the sense in which Weber used the term, and the other is in the sense in which Parsons employed the term. To Weber, power is an opportunity for an individual or a group of people to realize their own will through communal action, even against the resistance of others. Power in this sense is an aspect of social relationships, and hence power invariably means exercising power over others. According to Parsons, power is conceived of as a general facility or resource meant for the use or good of society. The more a society is able to mobilize it, the more power it will have at its disposal. In Parsons's frame of reference, power is seen primarily as a general capability to be used only by society.

In the context of tribes, what is implied by the term empowerment is something that has been going on for quite some time. I am interested here in locating the various forms that this process has taken and to understand the notion of power that these have entailed. I am also interested in the implications that these had for the tribes insofar as their relationships with non-tribes were concerned. Empowerment may be discussed both as a process and as a product. Insofar as tribes control and enjoy some power, the measures, however inadequate, may be conceived of as products. They may be seen also as processes insofar as they constitute stages or strategies in the larger process of the greater empowerment of the tribes.

Disempowerment Process

The issue of empowerment in India has arisen in two contexts. The first is the context of the traditional social structure as manifested in the caste system, which was both hierarchized and had a built-in structure of social inequality and social injustice. Disempowerment here is thus linked with the lack or absence of power. Empowerment here means conferment of certain rights that the groups did not have earlier. Conferment of rights on groups or communities that are ranked low in the caste hierarchy may be seen as evidence of the empowerment process (Béteille 1999). The second context in which the disempowerment process is located is not the context of the lack or absence of power, but the context of the appropriation and seizure of power. Here groups and communities held and enjoyed power, but this power was appropriated by groups or communities that came from the outside. Whereas the former was the case with caste society, the latter was more the case with tribal societies. Thus, in the case of tribes, the issue of their empowerment can be understood more meaningfully only in relation to the process of their disempowerment.

The process of disempowerment in the case of tribes began with the incorporation of the tribes into the larger social structure, especially the state. The incorporation was slow and gradual before the coming of the British. The consequences arising out of such a process were not intensely and violently felt or experienced by the tribes, although this did lead to the imposition of certain obligations on them. This meant that it was not incorporation as such that led to the process of disempowerment but the nature of the incorporation that took place. Where incorporation was marked by the seizure or appropriation of power by an agency from the outside, it invariably led to the disempowerment of the tribes. Disempowerment was of a higher or lower order depending on the nature of power at the disposal of the tribes. The incorporation that occurred during the British period was of this nature. The power held over the tribes and non-tribes by the British stemmed from the superior power—technological, military, and economic—that they held vis-à-vis the tribes. The disempowerment process was thus brought about, to begin with, by the use of brute physical force; there was nothing legitimate about it. This was checked during the later period of British rule, but it did not disappear entirely. The characteristic feature of domination and subjugation, enforced by the use of power, had far-reaching and unprecedented consequences for the tribes. This led to a strong reaction in some quarters. Elwin, for example, went to the extent of advocating a policy of complete isolation for the tribes. This appeared to him to be the only way of stopping the process of disempowerment that was at work during the period (Elwin 1944). Interestingly, in the post-independence period, until very recently, hardly any sensitivity was shown to the disempowerment processes at work among the tribes. Attention was paid to general social backwardness and low level of development of the tribes, given their geographical and physical isolation. The whole thrust of development policy in the decades after independence was to integrate tribes into the larger social structure or the so-called mainstream (Ray 1972). Hence, development of the tribes in the post-independence period moved along the same lines as those in the colonial period, with one exception—it was pursued much more aggressively in the period after independence. Correspondingly, the process of disempowerment that had been at work in the colonial period continued in the post-independence period. Not only that, the process also assumed enormous proportions after 1947, to which the Indian state itself contributed in no small measure.

This occurred mainly because of the imperatives of national and regional development. Tribal development could not be pursued outside the parameters of national development. Indeed, measures for bringing about rapid national development—such as building infrastructure, setting up industries, constructing dams and power projects—were seen as important steps for achieving the integration of tribal society. The national objective of building productive infrastructure for future growth and resource mobilization was seen as far more important than the issue of tribal welfare. Indeed, more often than not, the policy of national development was at odds with the policy of tribal interest and welfare, and the latter was invariably sacrificed in the name of national development.

The most important impact of the various programmes of national development was that they drastically altered the relationship of the tribes with the natural environment and its resources, namely land, forest, and minerals. Traditionally, these resources were either individually or collectively owned. The tribes at least had usufructuary rights over these. But because of the measures adopted by the state in the colonial period and later in the post-independence period, there has been a steady erosion of the control over and use of these resources by the tribes.

Land and forest are the resources that have affected the lives of tribal groups most deeply. The erosion of tribal land rights began under British rule and administration. It was brought about by a combination of forces at work during the colonial period. Of these, the two most important were the introduction of private property in land and the penetration of market forces in tribal areas. The two taken together opened up the way for large-scale alienation of land from tribes to non-tribes, especially after the tribal areas came to be linked with the larger society by roads and railways. The mechanisms through which this was achieved were fraud, deceit, coercion, and debt bondage, the latter being used most widely. Despite many protective and even restorative measures taken to stop land alienation in the post-independence period, there has been little success. Of course, the major source of land alienation in the post-independence period is not so much the encroachment of non-tribal people into tribal areas as the development policy followed by the Indian state. Large-scale industrialization, exploitation of mineral resources, and construction of irrigation dams and power projects in tribal areas during the period have uprooted more people from their lands than all the other factors put together.

Tribes experienced changes not only in their relationship to the land but also in their relationship to the forest. Tribes were greatly dependent on forests for their day-to-day needs, such as food, shelter, tools, medicine, and even clothing. But as long as the tribes were in control of forests, in the sense of having unrestricted use of forests and forest produce, they had no difficulty in meeting their needs. In turn, they preserved the forests, as this was their life-support system. The entry of the British drastically altered this relationship. To the British, forests were an important source of revenue and commercial gain. Hence, the forest policy adopted by the British introduced state control over forest resources and curtailed tribal rights and privileges over forest resources. The policy pursued by the British was continued in the post-independence period, focusing on economic development with even stricter regulation and enforcement. All these actions were justified on the ground that they were necessary for advancing wider and national public interest. In addition to affecting the rights of tribal communities, the forest policy and its enforcement had other consequences. It led to the steady depletion of forest areas. In 1854, the year that the first forest policy was adopted, the area under forests was estimated to be 40 per cent of the country's territory. In 1952, it was estimated to be only 22 per cent. Forest cover was estimated to have declined further to 10 per cent of the country's total area by the 1980s (Fernandes 1996).

As has been noted above, tribal areas witnessed large-scale industrialization, extraction of mineral resources, and development of infrastructure such as power projects, irrigation dams, roads, and railways. These have contributed to the increasing industrialization, urbanization, and economic diversification of tribal regions. Yet these developments have been of little benefit to the tribes. The industries and other development projects that have come up have not made jobs available to them. Even where they could have been absorbed, people from outside the region have been preferred over the tribal people. Tribes were thrown off their land and denied access to resources, leading to their displacement on a very large scale. Although this problem had begun to manifest itself in the first decade after independence, attention has been paid to only now. Further, benefits arising out of power and irrigation projects have not reached the tribes. Few tribal villages have been electrified, and tribal land under irrigation is almost negligible. Again, despite the urbanization of tribal areas, the percentage of tribal people living in urban areas is very small. In short, the fruits of development have not gone to the tribes but to people from outside. They find themselves increasingly subjected to poverty, exploitation, and oppression. In fact, these have become the marked features that characterize tribal communities in India, at least in regions outside the north-east. These being the general features of tribal societies at large, it is not surprising that they have come to be described as the fourth world in social science literature (Sengupta 1982).

Empowerment Process

The kind of concern expressed by Elwin during the British period finds more resonance today than ever before. But a way out of this problem today has not been thought of in terms of isolation but empowerment. Ironically, the very state that has been at the root of the disempowerment of tribes has also made certain provisions for their empowerment.

The Constitution, for example, contains many provisions for the benefit and welfare of tribal people. These constitutional provisions include creating scheduled tribal areas by bringing areas inhabited by tribes under the Fifth or Sixth Schedule of the Constitution according special treatment in the administration of tribal people; granting statutory recognition; providing proportional representation in Parliament and state legislatures; imposing restrictions on the right of ordinary citizens to move about freely or settle in particular areas or to acquire property in them; encouraging conservation of tribal languages, dialects, and culture; and extending special privileges in the form of the reservation of a certain percentage of posts in government services and seats in educational institutions. In addition, tribal people also enjoy protection under the fundamental rights meant for all citizens. Similarly, the Directive Principles of State Policy require that the educational and economic interests of the weaker sections of society, including tribes, be especially promoted. In short, the Constitution aims at safeguarding, protecting, and promoting the interests of tribal people.

It may be stated at the outset that there has been no formal statement of policy regarding tribes in India. Yet it would not be correct to say that there has been no state policy concerning tribes at all. A policy does exist, but it remains dormant and can only be inferred. Indeed, it is in the interest of the state that the policy remains unstated and nebulous. This gives the state an advantage in manipulating and taking measures that may be contrary to the welfare and interest of tribal people without being noticed or detected by the public. If one were to examine the constitutional provisions more carefully, one would see that the Constitution clearly adopts a position favouring tribal integration rather than isolation or assimilation, without using the term and concept of integration even once. Indeed, there is no official document where the policy of integration finds an explicit formulation. Yet this is what goes in the name of tribal policy in India. Of course, this is not without reason if one were to take into consideration the provisions made in the Constitution.

Integration as against assimilation (total fusion) and isolation (maintaining a distance) entails incorporation and acculturation into the larger society, but not at the cost of the tribes' own identity and distinct way of living. Yet there has been a big gap between theory and practice in the implementation of this policy. Policy and ideology stand at one end, and practice and programme at the other. If one were to examine the practice, one would discover that the policy was not one of integrating tribes but one of assimilating them into the so-called mainstream. It is indeed difficult to identify even a single programme under implementation that aims at fostering and promoting the distinct identity of tribal people in India except for those living in the north-eastern region. Nevertheless this fact, the idea of empowering tribes in theory could not altogether be ruled out. That tribes could not take advantage of the provisions in the Constitution and law or translate them into actual practice is a different story and constitutes the other side of the story of the empowerment or disempowerment process.

It is keeping in mind this backdrop that the state's agenda concerning the tribals needs to be examined and understood. The different measures adopted for the welfare of tribal people are usually divided into three categories. These are protective, mobilizational, and developmental. Protective measures include constitutional and legislative rights that safeguard the interests of the tribes (Roy Burman 1994). Mobilizational measures refer to the reservations extended to tribes in different fields. The developmental measures include programmes and activities aimed at promoting the welfare of tribal people. It is to this last aspect of the state agenda that scholars have paid the most attention.

Protective Measures

The process of tribal empowerment has a longer history than one may realize at first; it began during colonial rule. As noted earlier, the process of tribal empowerment is intimately related to the process of tribal disempowerment. Since the disempowerment process has become endemic in tribal society, any measure aimed at arresting such a process may, in a very general sense, be taken as evidence of the empowerment process. Some measures were negative in character, in that they did not confer any rights on the tribes but instead imposed restrictions on the rights of non-tribes. One such measure was the Scheduled District Act, 1874, in effect in certain parts of British India. Under this act, the operation of general acts and regulations in effect in other parts of British India was restricted in the areas described. The stated aim of this act was to protect tribals from the twin dangers of land alienation and indebtedness. In this way, tribes were protected from the greed of non-tribes. Some measures were positive in character, in the sense that they conferred rights on the tribes. For example, the Chhotanagpur Tenancy Act, 1908 and the Santhal Parganas Tenancy (Supplementary Provisions) Act, 1940 gave some rights to tribes aimed at protecting their land from being alienated.

Measures aimed at tribal protection and welfare during the colonial period were few and limited. However, as noted earlier, there has been a proliferation of such measures in post-independence India. Parliament and state legislatures have enacted further legislation in the same direction. Stricter laws have been adopted with a view to protecting tribal people. These rights, then, are seen as a sort of power granted to the tribes. Armed with this power, tribes are now able to protect and safeguard their interests. Most of the provisions for tribal protection adopted during the colonial period were of this nature. This has also been the most prevalent form of tribal empowerment in post-independence India.

No doubt, these rights gave tribes some power over non-tribes, particularly in relation to their land, by imposing restrictions on the power of non-tribes. It is a different matter that all tribes may not have had land. What is important is that these rights limited the scope of power enjoyed by non-tribes by treating it as illegal and therefore illegitimate. However, the grant of such rights has been of little use to the tribes as far as their actual enjoyment is concerned. These rights were aimed at helping tribes in exercising actual control over their life and resources. Yet this did not happen. The tribes continued to lose control over their land and other resources, a fact that can hardly be disputed. There were many reasons for the ineffective use of these powers by the tribes. To begin with, although the tribes were given rights, they were given these rights without any corresponding powers to enforce them. Although an agency to enforce these rights existed, enforcement was tardy mainly because of a lack of will on the part of the enforcement agency. Both the executive and judicial authorities were indifferent and insensitive to the rights of tribal people. In addition, the tribes themselves had very little say in these institutions. They continued to be dependent on the benevolence and zeal of sympathetic officials. Not only that, but there have been occasions when the rights conferred have been withdrawn under pressure from non-tribes. The legislatures that conferred such rights also made it a point to take away such rights when they found this to be inconvenient.

The Andhra Pradesh (Scheduled Areas) Land Transfer Regulation, 1959, as amended in 1970, for example, aimed at the prevention of land transfer from tribes to non-tribes. Any transfer of land under the regulation was to be considered null and void. However, this provision was withdrawn through a government notification in 1979 mainly under pressure from non-tribes. As per this notification, all non-tribals who were in occupation of lands in the scheduled areas up to 5 acres of wet land or 10 acres of dry land should not be evicted under the regulation of 1959. Thus, in spite of the act of restoration in effect since 1959, as many as 751,435.66 acres of land were still in the hands of non-tribals by 1995. Similarly, the Kerala Scheduled Tribe (Restriction on Transfer of Lands and Restoration of Alienated Lands) Act was included in the Ninth Schedule of the Constitution for fear of being challenged in a court of law. Yet the act has not gone unchallenged. The act was repealed by the Kerala Restriction on Transfer by and Restoration of Lands to Scheduled Tribes Bill, 1999. However, it failed to get the assent of the President. Following this, the state legislature passed the act under agricultural land, which comes under state jurisdiction of the constitution. The Kerala High Court nullified the act and the matter is now subjudiced (Bijoy 1999; Rao 1996; Verma 1990).

Mobilizational Measures

As noted above, one of the reasons why the rights could not be effectively enjoyed was because the tribes had very little say in the administration that affected their lives. One way in which this problem could have been overcome was by ensuring their involvement and participation in the organs of the state that were charged with the responsibility of implementing these rights. Yet it is not certain if it was believed that the presence of the tribes in the state institutions would help them in protecting their interests and rights. Of the constitutional provisions pertaining to reservation, two are of special significance. One is the provision of the reservation of seats in Parliament and state legislatures, and the other is the reservation of jobs for tribes in government and semi-government services. The extension of reservation provided an opportunity for tribes to participate in the decision-making processes of the state, both at the legislative and executive levels. They participated in the activities of Parliament and the state legislatures in enacting rules and regulations affecting society at large. All this was made possible by providing reservation to the tribes in the legislature and in government services. However, while this gave them access to state power, giving them a say in the decision-making processes of the state, they had very little say or power as far as their own communities were concerned. They had few rights in shaping or determining the life of their own communities. This was mainly for the following reasons. First, despite the provision of reservation, there were too few tribal representatives in the state institutions who could safeguard the interest and welfare of the tribes. Second, even when they were available, this did not necessarily give them access to power at levels where the interest and welfare of the tribes mattered the most. Third, because of both internal and external constraints, tribes had little room to exercise power or authority in the interest of their fellow tribes or communities.

The process of empowerment insofar as it gave tribes access to participation in the state structure also weakened the moral claim or authority that they had so far held on land and resources under their territories. Instead, it extended legitimacy to the state's claim for control over land and resources that the tribes themselves had earlier controlled and managed. Tribes were empowered by the endowment of one set of rights, namely share in state power, but were disempowered by losing their claim over land and resources that had been under their control so far. The state gave rights of one type but took away those of another kind. Indeed, in the absence of the extension of such rights to tribes, the state would have had no moral and legitimate claim over the lands and territories inhabited by the tribes and its resources. In fact, state action in the event would have seemed immoral, illegitimate, and unjustifiable. In the end, if one were to do a cost–benefit analysis of the gains made by the tribes, on the one hand and the state on the other the balance would surely go in favour of the state. In return for political empowerment (a share in the state decision-making process) that the state extended to tribes, it took away economic empowerment that they had held so far over the territories in which they lived.

Reservation of seats in Parliament and state legislatures and reservation of jobs in government services, especially for the unprivileged, has generally been seen as one of the important aspects of the general process of empowerment. Since empowerment entails some dimension of power, as noted above, one is tempted to ask questions about the notion of power that may be associated with the provision of reservation. No doubt, it extends some rights—and thereby some power—to the unprivileged. In this instance, it is not a case of giving power to the tribes at the cost of the non-tribes. Rather, it is an attempt to ensure the effective distribution of state power among citizens. Given this, the notion of power entailed in the case is far from clear. It certainly does not fall within the Weberian framework. In the present context, there is hardly a case of power being exercised over others. Tribes, for example, do not hold power at the expense of others. Hence, empowerment through means such as reservation does not necessarily mean a loss of power for the non-tribes. In this context, power may thus more meaningfully be understood as a facility or resource placed at the service of the tribes.

Developmental Measures

The other kinds of measures adopted by the state for tribes were developmental. Such measures had, of course, nothing to do with power or rights. All the same, the gains made under such measures could have gone a long way towards the realization of these rights. However, these measures contributed little in that direction. This was because all constitutional and legal measures for empowerment, and those that existed at the level of protection and mobilization, were thrown to the winds. Hence, the development policy pursued by the state led to the increasing disempowerment of the tribes in terms of access to resources that they had so far enjoyed. The notion of undertaking development as part of Panchshila remained only an idea. The principles underlying Panchshila as enunciated by Nehru were the following. People should develop along the lines of their own genius, and we should avoid imposing anything on them. Tribal rights in land and forest should be respected. We should try to train and build a team of tribal people to do the work of administration and development and avoid introducing too many outsiders into tribal territories. We should work with and through, and not in rivalry with or against, their social and cultural institutions. We should judge results not on the basis of the amount of money spent but on the basis of the quality of human character evolved (Elwin 1960). Unfortunately, such a philosophy of development for the tribal people was never translated into reality except in a limited way in the north-eastern region. In the region, Panchshila found concrete expression in the institution of the district council to begin with and with the state structure at a later stage.

Institutional Structure

In view of the above, empowerment in the context of tribes has assumed new meaning today. It is true that empowerment is being talked about increasingly in the sense of granting more and more power to tribes. Yet empowerment is not being talked in the sense of more legal rights or reservation in state institutions, but in the sense of institutional mechanisms or granting power to the people whereby they are able to exercise greater control over their own lives. In the context of the tribes, this means power to control their own resources and to determine and regulate their own lives according to their own needs. This means power may be vested either in traditional social institutions or in new institutions created specifically for this purpose. Empowerment here entails power in both the Weberian and the Parsonian senses.

Empowerment even in this sense of the term is not new. One can find traces of this in the Indian Constitution. It may be recalled that there is a provision of the Sixth Schedule areas in the Indian Constitution. The Sixth Schedule contains provisions in relation to the administration of tribal areas. Tribal areas in the Constitution meant regions inhabited by tribes in the state of Assam. Today, most of this region has been carved into separate states, indicating thereby the further process of empowerment of the tribes.

Autonomous District/Regional Council

The provisions under the Sixth Schedule purport to provide a self-contained code for the governance of tribes living in the Sixth Schedule areas. The avowed objective of the Sixth Schedule is to ensure the grant of autonomy and self-management rights to the tribes. In short, it provides for the institutions of the autonomous district and the autonomous regional councils. The scheme of administration was almost wholly based on the recommendations of the North-East Frontier Tribal and the Excluded Areas Sub-Committee of the Constituent Assembly. The committee was set up to report to the Constituent Assembly on the scheme of administration of the tribal areas. The scheme that was conceived was based on the need for establishing an autonomous administration in these areas so that tribal people could continue to follow their traditional ways of life and to introduce changes that they themselves desired.

The separate scheme of administration was accepted on three considerations. First, the social customs and social organizations of tribal people were distinct from those of non-tribal people. Second, it was feared that tribal people would be exploited by non-tribal people on account of the latter's superior organization and experience in business. Third, it was also feared that unless suitable financial provisions were adopted or powers conferred on local committees, the provincial government would not set apart adequate funds for the development of tribal areas under pressure from the plains people (JHADC 1997).

The councils have thus been given powers equivalent to those exercised by a state or union. They have three wings—legislative, executive, and judicial. They have been given powers to make laws in matters connected with social customs, land, forests, etc. They have also been given judicial and regulatory powers in such matters as control of money-lending and trade by non-tribals and establishment of village and town committees. They also enjoy executive powers in such matters as establishing and managing primary schools, dispensaries, markets, fisheries, ferries, village roads, bridges, footpaths, playgrounds, community halls, etc. They have also initiated measures for better health, medical, and sanitation facilities for the areas under their jurisdiction (Roy Burman 1997; JHADC 1997). In view of the powers vested in the councils, tribes in the north-east have been able to protect themselves from the exploitation of non-tribals to a far greater extent than the tribes in other regions.

Creation of New States

Despite the fact that the autonomous district council granted some powers to tribes determining many aspects of their life, these powers suffered from certain limits. There were two reasons for the limited powers of the district council. The first reason was the limited powers given to the district council in the first place. We have already discussed how power was given to the district autonomous council. The second—and main—cause of the limited powers of the district council was the actual politics of the state of which the tribal regions are a part. State-level politics was characterized by a deep divide between tribes and non-tribes, or between the plains people and the hills people. As noted above, state-level politics was dominated by the non-tribes, who were numerically dominant and socially and economically powerful. Not only did they tend to dominate the tribes but the politics they followed also tended to thwart and scuttle the functioning of the autonomous district council by blocking grants and creating other impediments (Roy Burman 1997).

The district councils thus experienced a variety of difficulties in the exercise of the power conferred on them. To begin with, law-making power in general was constrained by the fact that until assent was given by the governor, the law promulgated by the district council could have no effect. At the same time, while the district council demanded that the governor should exercise his discretion in the matter, judicial pronouncements suggest that the governor is to be guided by the advice of the ministry. Only through a recent constitutional amendment has an exception been made in respect of the district councils of Karbi Anlong and the North Cachar Hills. Over and above this, the power to make rules is expressly limited by the provisions of the Sixth Schedule that has created them, and, therefore, they can do nothing beyond the powers given to them. The district council can exercise executive functions in respect of several matters listed in the Sixth Schedule. Yet it is not mandatory under the Sixth Schedule to vest any development function with the district council. At best, the concerned state government may, with the consent of the district council, assign development functions to it. However, the state government has not been as liberal in making the district council functionless or in putting it under suspension, initially for six months but usually for the period to be extended by fresh notification. The sources of revenue of the district council or of the regional council have been specified in the Constitution, but some major sources of revenue are under the control of the state government. The latter is expected to pass on the share of the council after due computation, which sometimes takes years. In Assam, the district councils today enjoy the privilege of being represented at the discussions of the Planning Commission along with the government of Assam. They obtain specific allocations and funds from the Planning Commission. But these funds have to be routed through the government of Assam. It is alleged that even these funds are not released in time. Given the fact that many state governments are in the habit of drawing an overdraft, one should not be surprised at the temporary or even the permanent diversion of funds by them. An important fact to note is that some of the revenue-raising powers of the autonomous district council in the Sixth Schedule areas are less than the powers actually granted to even the gram panchayat in the non-tribal areas of some states. Thus, the role and functions of the district council have fallen short of the expectations of the tribal people in the region (Roy Burman 1997). The demand for a separate state emerged out of this situation. The creation of new states for the tribes was the culmination of the demand for greater empowerment of tribal people.

Gram Sabha

This demand for greater empowerment has spread to other parts of tribal India and has led to the creation of new institutions, such as an autonomous regional council for the Bodos as well as for the tribes of Jharkhand. It has also taken the form of the revival of old institutions such as the gram sabha. Indeed, the power, control, and jurisdiction exercised by the gram sabha earlier have now been given legal recognition by an act of Parliament. It may be recalled that the 73rd and 74th Amendments to the Constitution stipulated that the provisions contained therein will not apply automatically to the scheduled areas covered by the Fifth and Sixth Schedules. These were to apply only with such modifications and amendments as Parliament may decide. As a follow-up to this provision, the Ministry of Rural Development, Government of India set up a high-level committee to recommend the contents of a parliamentary act for the Fifth Schedule areas. Based on the report of the high-level committee, the Panchayat Extension to the Scheduled Areas Act (PESA), 1996 was passed by Parliament in December, 1996.

The most important provision made in this act is that state legislation on the panchayat would be in consonance with customary law, social and religious practices, and traditional management of community resources. This means that the village would ordinarily correspond with the community, managing its affairs in accordance with traditions and customs. Every village would have a gram sabha, consisting of persons whose names are included in the electoral rolls for the panchayat at the village level. The gram sabha is vested with the power to safeguard and preserve the traditions and customs of the people, their cultural identities and community resources, and to uphold the customary mode of dispute resolution. It is also vested with the power to approve plans, programmes, and projects for social and economic development before these are taken up for implementation by the panchayat at the village level. The gram sabha is to identify or select persons as beneficiaries under poverty-alleviation and other programmes. Even the panchayat at the village level will be required to obtain from the gram sabha a certification of utilization of funds by that panchayat. Further, the gram sabha or the panchayat at the appropriate level would be consulted before acquiring land in the scheduled areas for development projects and before settling or rehabilitating persons affected by such projects. Again, the recommendations of the gram sabha or the panchayat at the appropriate level would be mandatory prior to the grant of prospecting licenses or mining leases for minerals or the grant of concessions for the exploitation of minor minerals by auction. The state government is expected to endow the gram sabha with the power to take up various activities. These include enforcement of prohibition, regulation or restriction of the sale and consumption of intoxicants, regulation of ownership of minor forest produce, prevention of land alienation and restoration of any unlawfully alienated land, management of village markets, control of money-lending, and supervision of plans and resources as well as institutions and functionaries. In short, for the gram sabha to function as an institution of self-government, the state government is required to ensure that the panchayat at the higher level does not assume the power and authority of any panchayat at the lower level, namely the gram sabha (Government of India 1996).

The Problem of Democracy and Freedom

The creation of the institutional structures referred to above represented an incremental advance over other forms of tribal empowerment as it enlarged the scope of tribal power in two ways. First, the tribes were given some legislative, executive, and judicial powers that they did not have in other forms of the empowerment process. It is a different story that the issues over which they held such powers were very limited. The institution of the district council may be taken as an apt case to illustrate this point. Second, the tribes were given more powers over a wider range of issues than earlier. The creation of state structures for the tribes is illustrative of this. Empowerment here implies the enlargement of power not only in the sense of power over a wider range of issues than had been granted earlier, but also in the sense of the nature of the power conferred and enjoyed. The tribes, for example, enjoyed more power in the state structures than they did in the other two institutions, namely the district council and the regional council. In addition, with the creation of such structures, tribes came to enjoy more effective control of power. That is, there was a relatively more effective distribution of power in favour of tribes.

However, even when there has been empowerment of tribes as opposed to non-tribes, as in the case of the district or regional councils earlier and the creation of new states in recent years, the benefits of empowerment have not percolated to the people at large. To begin with, there has been a general tendency at the state level to monopolize power rather than share it with people at large. The law-making power of the district councils in Meghalaya, for example, invariably experienced problems with the principle of repugnance. Any law or regulation promulgated by the district councils if repugnant to any law made by the state legislature was to be null and void, and the law enacted by the state legislature was to prevail. This made a mockery of the power and autonomy granted to the district councils. Thus, while the creation of certain state institutions gave more power to the tribes, it also led to processes that arrested the exercise of the power and autonomy at the grassroots level. In other words, the grassroots democratic political process was reduced to a political toy in the hands of the power elites, bureaucrats, and politicians at the state level (Roy Burman 1997). It is rather paradoxical that the state government manned by the tribes should come in the way of the smooth functioning of the district council. There is, then, a tendency within tribal society towards a greater monopolization of power by the elite sections of that society.

The empowering of tribes is primarily against non-tribes. At the same time, it is worth bearing in mind that there is hardly any territory, region, or district that is inhabited exclusively by a single tribal group. Where more than one tribal group exists, there invariably exists a certain degree of inequality and domination along tribal lines. Certain tribes are not only likely to be numerically dominant but are also likely to be economically and politically more powerful than others. In such a situation, the empowerment of tribes may not necessarily mean the same thing to all the tribes living in a given territory. Some may stand in a more advantageous position vis-à-vis other tribal groups, especially the smaller ones, and are likely to exercise domination and power over others. Hence, empowerment may not necessarily lead to the empowerment of all tribal groups in a given region or territory. The empowerment of tribes may even fall short of the ideals of equality and social justice as embodied in the concept of citizenship. There is no doubt that tribal societies have relatively more egalitarian in comparison to non-tribal societies, but they are not altogether free of inequality. Besides the inequality inherent in the traditional social structure, especially inequality based on gender and lineage, such inequality in tribal society is increasing because of uneven access to the benefits of modernization. Hence, even with empowerment, marginalized sections such as women and commoners may remain without any power unless the notion of citizenship is also instilled in tribal society.

References

Béteille, Andrè. 1999. 'Empowerment: The Indian Way.' *Economic and Political Weekly* 34 (22): 5892–97.

Bijoy, C. R. 1999. 'Adivasis Betrayed: Adivasi Land Rights in Kerala.' *Economic and Political Weekly* 34 (22): 1329–35.

Elwin, Verrier. 1944. *The Aboriginals*. Oxford pamphlets on Indian Affairs, no. 14. Bombay: Oxford University Press.

———. 1960. *A Philosophy for NEFA*. Shillong.

Fernandes, Walter. 1996. 'The New Forest Act: Conservation Orientation But Where Are the People?' (mimeograph).

Government of India, Ministry of Law and Justice. 1996. The Provision of the Panchayats (Extension to the Scheduled Areas) Act.

Jaintia Hills Autonomous District Council (JHADC), Meghalaya 1997. Seminar paper on Autonomy and Self-governance (mimeograph). Developing Countries Research Centre (DCRC), Delhi University.

Rao, B. Janardhan. 1996. Tribal Land Rights, Government Action, and Socio-political Movements (mimeograph). Rajiv Gandhi Institute for Contemporary Studies, New Delhi.

Ray, Niharranjan. 1972. 'Introductory Address', in *Tribal Situation in India: Proceedings of a Seminar*, edited by K. S. Singh, Shimla: Indian Institute of Advanced Study, pp. 3–24.

Roy Burman, B. K. 1994. *Tribes in Perspective*. Delhi: Mittal Publications.

———. 1997. Sixth Schedule and Seventy-Third and Seventy-Fourth Amendments of the Constitution. Keynote Address in the Workshop on Decentralization and Self-Governance: Experiences of a District Council in North-East India (mimeograph). Developing Countries Research Centre (DCRC), Delhi University.

Sengupta, Nirmal. 1982. *Fourth World Dynamics: Jharkhand*. Delhi: Authors Guild Publications.

Verma, R. C. 1990. *Indian Tribes Through the Ages*. New Delhi: Publications Division, Ministry of Information and Broadcasting, Government of India.

Chapter 7

The Politics of Language, Religion, and Identity: Articulation of Social Difference

Language and region, each on its own, are important marks of difference, but they tend to coincide with each other in the Indian context. In fact, together have been the driving force behind the reorganization of society and polity in India in the post-independence era. Language and region, though important, however, do not tell us anything about the inner workings of the society and polity that they embody. What is even more striking is that despite differences based on language and region, societies are enormously similar in their societal characteristics—religion and caste being predominant among them. Indeed, religion and caste cut across region and language, and to that extent constitute the common threads across the diversity of languages and regions (Dumont 1998). It is this commonality that has come to be the principal unit of sociological study and discourse in India. This partly explains why language and region have not become rallying points for sociological deliberation and analysis. Analyzes of language and region are not unimportant, but these aspects were embedded in the institutions of caste and religion. It is within this commonality of Indian society that tribes in India have been posited and are constituted by sociologists and social anthropologists in India.

The commonality has been so pervasive and persuasive that even colonial ethnographers could not escape its influence when they began conceptualizing tribes in the Indian context. Thus, to colonial administrators-cum-ethnographers, tribes came to be constituted as peoples who practiced animism or tribal religion. In such a conceptualization, colonial administrators described tribes as communities located outside of historical and textual religions and their accompanying social organizations. Of course, other criteria such as primitive living conditions and living in isolation were not left out of the analysis, but they followed more as a corollary from the first postulate.

In anthropological literature, tribes in general have been defined in terms of language, culture, territory, and government. However, these features have not been brought to the centrestage of sociological discussion. Béteille in his essay (1960) has discussed these issues and pointed to the kinds of inadequacies that these present in the Indian context. However, a new analysis based on these features as a starting point was not pursued. As stated earlier, tribes came to be conceptualized residually, in terms of the contrast with the general or universal features of Indian society rather than the particular features they embodied. This is not to say that the particular features were overlooked, but that they were not at the centrestage of the conceptualization of tribes. For people at the grassroots level, it was not so much the common features, namely caste and religion that mattered. Rather what mattered most were language and culture, now often referred to as ethnicity, and were considered the most pronounced markers of distinctiveness. In the regional context, tribes were invariably posited against the dominant regional community, which also happened to be a distinct linguistic and cultural community. This is evident from the fact that tribals are addressed by their ethnic or tribe names, which generally correspond with their distinct language or dialect. Yet this aspect of the labeling of tribals has been overlooked in the sociological discourse on tribes. In the attempt to differentiate on the basis of general or universal features (albeit in a limited sense), what has been overlooked are the differences on the basis of which people themselves differentiated between one another.

Recognition of Differences as Types: The Social Science Perspective

That there is a difference between tribes and larger Indian society is widely accepted. Indeed, the difference has been taken as the starting point for understanding the growth and expansion of Indian society and the nature of social transformation in tribal society. However, sociologists and social anthropologists do not agree on the ways of conceptualizing the difference. The overwhelming majority sees the difference as one of structure. In this view, the difference between tribes and the larger Indian society is one of type and not of kind. Thus, tribes have invariably been seen as lying outside the larger Indian society. To put it plainly, tribes were seen as not being part of civilization, and, therefore, were considered outside the structure of the larger Indian society. And yet they were seen not as isolates but as being in constant interaction with so-called civilization, at least in the case of those on the fringes of the larger Indian society. Hence, tribal societies were seen as undergoing change because of the process of acculturation resulting from their contact and interaction with larger Indian society. This process of acculturation has been viewed in terms of tribes becoming a part of larger Indian society. Thus, Kosambi (1975) refers to the growth and expansion of Indian society resulting from a fusion of elements belonging to tribes and larger Indian society. Bose (1941) points to the process of how tribes were drawn into the social organization of the production system of the larger Indian society, which invariably provided protection and security. This was how, according to him, tribes were absorbed into the larger Indian society. He called the process 'the Hindu method of tribal absorption'. Srinivas (1977) talks of the process of Sanskritization through which the lower-ranking castes emulated the way of life of the higher-ranking castes. Although the concept of Sanskritization has been used mainly to understand the process of social mobility within the caste hierarchy, it has also been extended to understand the process of transformation in the context of tribal society. Sinha (1962) also points to the processes of Sanskritization and Hinduization, but, according to him, these take place invariably within the framework of state formation.

Now in this understanding of Indian society and its social transformation, tribes as entities with distinct languages, cultures, customs, traditions, and, above all, social organizations, are at least recognized. Indeed, these differences act as the points of departure in the analysis and understanding of the transformation of tribal societies. They point to how as tribes come in contact with the larger Indian society, they tend to lose their distinct identities and are absorbed into the latter. According to this line of thinking, this process of social transformation leaves tribes with no distinct identity and space of their own. If at all they have an identity, this is an identity only as a caste within the overall organizational and hierarchical structure of the caste system.

Denial of Differences as Types: G. S. Ghurye and Right-Wing Political Thinking

According to another school of thought, the difference between tribes and non-tribes, as recognized by the scholars mentioned earlier, has been overlooked. This is not to say that the difference was not recognized. Rather, what was not recognized was the difference of a type as articulated by most social anthropologists. The origin of this line of thinking can be traced to G. S. Ghurye's writings on tribes. Ghurye in his book, *The Aborigines—"so called—and their Future*, argued in favour of describing tribals as Hindus; the expression he used was 'backward Hindus'. The book has been issued several times since it was originally published in 1943 under a new title, *The Scheduled Tribes*. Ghurye divides tribes in India into three groups: those that are properly integrated; those that are loosely integrated; and those that are not more than touched by Hinduism. Referring to the last group, he says that the only proper description of these people is that they are imperfectly integrated classes of Hindu society and that they are in reality backward Hindus. He makes this argument on the grounds that there was much similarity between Hinduism and the animistic tribal religions and that the two could not possibly be distinguished from one another. This argument was based not on fieldwork but on the observations and comments of some of the Census Commissioners between 1891 and 1931, where they expressed their dissatisfaction with the fact that tribes were described as animists. It is, of course, an open secret that Ghurye made his observation based on highly inadequate data and on the very selective use of the comments and observations of the Census Commissioners (Ghurye 1963: 1–22).

Indeed, Ghurye's conceptualization of tribes as 'backward Hindus' set the tone for a future line of thinking on tribes in India. It found a very strong resonance in the report of the Christian Missionary Activities Enquiry Committee. The committee, more popularly known as the Niyogi Committee after its chairman, was constituted to look into the activities of Christian missionaries, tracking to the conversion to Christianity by tribes in the erstwhile undivided state of Madhya Pradesh. The committee cited extracts from Ghurye's book in support of its own recommendations (Niyogi 1956: 27–29). Since then, the view that tribal people are Hindus has become a popular refrain among right-wing Hindu social and political activists. The position is reiterated and articulated aggressively by the Sangh Parivar in its discourse on tribes in India. Indeed, this has become the pattern of thinking about tribes in India today.[1]

By bringing religion to the forefront of the conception of tribe, the Sangh Parivar and other right-wing ideologues show a marked continuity with the colonial tradition. Yet there is also a departure in a certain respect from that tradition. In the colonial literature, tribes were no doubt characterized on the basis of their distinctive religion but they were also described on the basis of other criteria, especially their isolation from the larger society. According to the latter view, tribes were regarded as a society unto themselves and hence as constituting a different society vis-à-vis the larger society. What is new as far as the advocates of Hindutva are concerned is that they have begun to conceive of tribes solely in terms of religion. Yet even here the proponents of Hindutva adopt a position different from the colonial point of view.

According to the colonial tradition, tribes were classified as animists, a pointer that they belonged to a religious tradition other than that of the major religions of India. The advocates of Hindutva, however, conveniently overlook this fact and categorize them as Hindus. Indeed, they have gone even further. In keeping with their new interpretation of the concept of tribes, they have begun denying the status of tribes to those who otherwise belong to the same ethnic group, speak the same language, share the same social organization, but who practice and adhere to a tradition other than the Hindu religious tradition. Such an articulation, hitherto dormant and latent, has now come to the surface, as seen in the recent attacks on tribal Christians in India. As the media coverage of these attacks reveals, the Sangh Parivar activists have time and again forcefully and aggressively declared that tribes cease to be tribes once they became Christians. They also insist that tribes must identify themselves as Christians and not as tribes when they apply for jobs and other government benefits, on the grounds that when they become Christians, they cease to be tribes and are therefore ineligible to apply for state benefits as tribes.[2]

Now such a conception not only goes against the general anthropological understanding of tribe but it also militates against the basic conception and spirit underlying the Constitution of India. Certain groups or communities have been identified as tribes not because they follow a particular religion but because they have been enumerated as scheduled tribes under the Indian Constitution. These groups or communities are Scheduled as Tribes not because they practice a particular religion but because they constitute a particular community distinct from the dominant regional community. They generally speak their own languages and have a distinct social organization and way of life, quite different from that of the dominant regional community.

Anthropological Reading Versus Politico-Administrative Practices

It is a paradox that the historical and anthropological understanding of tribes dominant in social science literature has had little influence on popular thinking about tribes in the post-independence period. Indeed, political and administrative practices concerning tribes are influenced more by the general thinking than thinking stemming from social science inquiry. The dominant social science thinking on tribes in one and the dominant political and administrative thinking and practices in another. This is contrary to the way anthropology had functioned under colonialism. Anthropology had then been a tool in the pursuit of colonial interests and hence was the handmaiden of the colonial state. In the postcolonial period, however, the role of anthropological writings in shaping state policy and strategy pertaining to tribes, despite the presence in the state administration of noted anthropologists as advisors or administrators, has been almost negligible unless it has been convenient to the dominant political thinking about tribes. State policies and administrative practices pertaining to tribes in matters of language, culture, and religion are, for example, contrary to anthropological understandings of them. These issues have been taken up for discussion later.

Challenges to Social Science Concepts and Theories

The thinking enunciated by Ghurye and advocated aggressively by the Sangh Parivar has far-reaching implications for the way in which tribes have been discussed and understood in social science inquiry so far. That this will also have a significant bearing on the future course of policies and programmes on the tribal question in India can hardly be denied.

The categorization of tribes as Hindus leads to conceptual and empirical difficulties. To begin with, whether tribes are to be treated as Hindus is a debatable question. There are both similarities and differences in the religious practices of Hindus and tribes. The protagonists of Hindutva have, however, conveniently overlooked the differences. Even on the matter of similarities, it is not tenable to treat tribes as Hindus. Similarities have been drawn based on two sources of evidence. The first is the influence of Hinduism on tribes, and the second is a similarity owing to the fact that both are, to a greater or lesser extent, natural religions (Bose

1941; Elwin 1960; Ghurye 1963).[3] There is no doubt that there has been much give and take between the two religions. However, the influence of Hinduism on tribes, though present, is not adequate grounds for describing tribes as Hindus. The second aspect that has been alluded to is the dimension of natural religion. As a natural religion, tribal religion shares many attributes in common with Hinduism, as it does with the religious practices of tribes in the Americas and in Africa.[4] Yet it is doubtful whether the religious practices of tribes in the Americas or in Africa can be described as Hinduism, or whether those tribes can be alluded to as Hindus. To categorize tribes in India as Hindus, therefore, smacks of cultural and religious expansionism. Just because there are some similarities, the tribes cannot be denied their distinct identity and autonomy. It is important to note that religion does not include only practices. It also includes ideas and beliefs. However, this fact is completely glossed over even by sociologists and social anthropologists in their discussions and analyzes of tribal religions. The fact that tribes have their own world view is not granted any place or recognition in such deliberations.[5]

Second, if tribes are to be regarded as Hindus, then the entire historical process depicted by historians to explain the evolution and development of Indian civilization is open to contestation, and even rejection. The same would be the case with the conceptual apparatus of Hinduization, acculturation, assimilation, and absorption developed and used to understand the dynamics of Indian society. Hinduism is intricately linked with the caste system, and it is not so much against religion as against caste that the social organization of tribes has generally been posited in social science literature. This makes it impossible for a tribal to be a Hindu and a member of a tribe at the same time. One can be a Hindu only at the risk of losing one's tribal status. The two cannot go together. One can, of course, acquire a new status, but that new status is of caste rather than that of tribe. That is how social scientists, including historians and anthropologists, have viewed social change among tribes (Bose 1941; Sinha 1962, 1987; Kosambi 1975). The implication is that in social science literature, tribes and Hindus are two incompatible entities, at least as the point of departure for analysis and understanding. However, it is precisely this view that the votaries of right-wing political thinking have been questioning. Social science inquiry, on the one hand, and the dominant political thinking on the other, seem to be at loggerheads with each other in their conceptions and understandings of tribe. While this is so, it is important to note that social science literature does not deny the process of Hinduization and the eventual transformation of tribal people into Hindus. However, this process is viewed as the end-point and not as the starting point, in Ghurye's formulation (1963).

Constitutional Provisions: Recognition of Differences

These difference in the conceptualization of tribes gave rise to two conflicting views regarding the policy to be adopted towards them in India. These have been popularly identified as the policies of isolation and assimilation. The former is attributed to Elwin, though many colonial administrator–ethnographers also largely shared it. The latter is associated with Ghurye. The two positions were, of course, connected with the conception of tribes that was based on, but did not exclusively draw on, that understanding. Some anthropologists shared Elwin's conception of tribes but did not necessarily adopt the same line of argument that he did. Nor did they support or share Ghurye's viewpoint. It is generally believed that the two anthropologists subsequently abandoned their respective positions and argued more in favour of integration than isolation or assimilation. It is somewhat puzzling that Ghurye should have advocated the policy of assimilation given the conception he held of tribes. To argue that tribes were Hindus, as he advocated, meant that tribes were an integral part of the larger society. In a sense, then, the advocacy of the policy of assimilation seemed redundant. Ghurye's stance was more of a ploy to counter the position held by Elwin.

In the first few decades after independence, deliberations on the tribal question among scholars, administrators, social workers, and politicians focused on the need to facilitate the integration of tribes into the larger society. Since then, leaders, planners, academicians, and bureaucrats at the national level have been talking of the state policy concerning tribes as one of integration. It is a different matter that there exists no formal statement or document in regard to the policy to be adopted towards tribes in post-independence India.

The state's policies and attitudes towards tribes could, however, be discerned from the kinds of provisions laid down for tribes in the Indian Constitution. The special provisions for tribes include statutory recognition, proportional representation in legislatures, the right to use their own language for educational and other purposes, the right to profess their own faith or freedom of faith, and the the right to pursue development—economic and social—according to their own genius. The Constitution also empowers the state to make provisions for reservation for tribes in the areas of jobs and appointments. The Constitution also empowers the state to bring areas inhabited by tribes under the Fifth or the Sixth Schedule for purposes of special treatment in respect of the administration of tribal people. In addition, the Directive Principles of State Policy of the Constitution require that the educational and economic interests of the weaker sections of society, including tribals, be especially promoted.

Now if one carefully reads the special provisions contained in the Constitution for the tribes, one finds that these provisions reveal an approach that may be described sociologically by the term 'integration'. The aim of the constitutional provisions pertaining to reservation in education, employment, political representation, and administration of tribal areas as well as the provisions laid down in the Directive Principles of State Policy is to bring the tribes closer mainstream Indian society. The Constitution also has provisions for not only protecting and safeguarding the languages, cultures, and traditions of the tribes, but also for promoting them. In this sense, the overall thrust of the Indian Constitution concerning the tribes is one of integration rather than assimilation. Although the meaning of integration may be understood in different ways, the underlying assumption is that it provides space for diversity, unlike the policy of assimilation, which extends no such space. The conception of assimilation entails that the small minority must give up its culture in favour of the culture of the dominant majority. It is a paradox that notwithstanding the constitutional provisions, no official document exists where the policy of integration is explicitly stated or discussed. Yet integration is often cited or upheld as the tribal policy in India. Often national leaders, planners, and administrators take up the principles that Nehru had enunciated in his foreword to Elwin's book, *A Philosophy for NEFA*, 'as [a] statement of the national policy towards tribes' (Elwin 1960). It is intriguing that a national policy has to be sought in a book rather than in an official government document. The ambiguity of policy is, of course, advantageous for the state. The state can deploy it as and when it finds it convenient. Further, it is doubtful if the principles enunciated by Nehru could be taken as the national principles. After all, these principles never formed the guiding spirit of tribal development for mainland India. They were at best guidelines for development for north-eastern India, especially the Sixth Schedule areas and the erstwhile NEFA. Indeed, it would not be out of place if one were to treat Nehru's principles as nothing more than the explication of the rationale underlying the provisions in the Sixth Schedule of the Constitution.

Rupture Between Constitutional Provisions and Administrative and Political Practices

What is stated in the Constitution, laws, and statutes is one thing and what actually happens is quite another. Often, there are gaps between what is stipulated and promised and what is actually done. Such gaps are more pronounced in situations where the beneficiaries happen to be small, backward, inarticulate, and dominated, as has been the case with tribals in India. Their future is, therefore, contingent on the benevolence of the larger Indian society, especially its leadership. Often, however, the leadership has not shown such benevolence. After all, tribes have never been placed on the same footing or been seen in the same light as members of the larger society in terms of their access to rights, whether civil, political, or social. They have invariably been seen as those who have had to make place for the larger society and the agencies that protect their interests, namely the state.

Despite the proclamation of lofty ideals and the adoption of various constitutional and legal provisions for the protection and welfare of tribal people, actual developments have moved in a direction opposite to that which has been promised. Indeed, the achievements of the state in this context have been more in the direction of extending civil and political rights to tribes vis-à-vis the larger society to which they have been drawn. The gains have thus been more in the direction of bringing the tribes closer to the state and to the larger Indian society through measures such as reservation in politics, education, and government employment. However, the lack of success has been most striking precisely in the domain aimed at protecting their culture and tradition, so crucial to the ethos of integration described by scholars and administrators as underlying the national tribal policy. The protection of tribal lands—their life-support system—has been poorly implemented, but at least there has been some symbolic effort in this regard both in terms of legislation and implementation (Singh 1986).

The Constitution recognizes the distinct cultural features of tribes, especially with respect to language, and talks of their protection and promotion. At the same time, as part of the process of integration into the larger Indian society and polity, tribes have been given citizenship rights. As citizens, tribes enjoy fundamental and civil rights such as the right to freedom of expression and freedom of faith. Yet the reality has often been different, for they have been denied such rights in practice, as can be seen in the area of tribal language and religion. According to Article 29 of the Constitution, a cultural or linguistic minority has a right to conserve its language and culture. The article provides protection to scheduled tribe communities to preserve their languages, dialects, and cultures. It also says that the state can not by law enforce upon them any other culture or language. Further, Article 350a provides for facilities for instruction through the mother tongue at the primary stage of education. Yet no effort whatsoever has been made so far by the central, state, and local governments to safeguard tribal languages, much less their promotion. Rather, education in all states and union territories, even at the primary level, has been imparted in the language of the dominant community. In some states like undivided Bihar, primers were prepared in some tribal languages for pedagogical purposes in the mother tongue, but these were allowed to rot in government godowns.

One may argue that since the Constitution has given the tribes certain rights, it is left to the tribes themselves to take care of their culture. As far as the state is concerned, it will not come in the way, and it may even give assistance in the form of grants and infrastructural facilities. If this is indeed the case, it is perhaps too much to expect this from communities with scarcely any literacy. At the same time, it is worth noting that governments have been far from neutral in regard to the issue of language. The dominant regional languages have been actively supported and promoted by respective state governments. However, there is no level playing field for the promotion or protection of tribal languages despite the constitutional provisions. At the same time, the tendency in regional politics has been to give tribes the regional identity of the dominant-language-speaking people, which often weakens or threatens their cultural identity. Looking at the access to resources and facilities available to the tribes, whether in politics or employment or education, we see that they are invariably excluded on the grounds that they belong to a different ethnicity. In short, in certain contexts, or for certain purposes, the emphasis is on absorption, which entails the inclusion of the tribes into the dominant society while in others the difference is not only maintained but also covertly employed to deny, to discriminate, and even to prevent them from gaining access to wider social and cultural relationships. What obtains is a strategic deployment of difference.

The domain of religion is another area where the distinctiveness of tribes is denied by the state. As stated earlier, one of the important criteria used for the delineation of communities as tribes was religion. Generally, tribes were identified and delineated vis-à-vis larger Indian society, which practiced Hinduism primarily, but contained were adherents of other religions as well. In contrast, tribes were described as those groups that practiced animism. That is how tribes were differentiated from the rest of the population. However, those who practiced animism also represented different kinds of social organization and different languages, customs, traditions, and social practices. These were implicit and not explicitly articulated in the delineation of tribes.

Since the inception of separate religious electorates in 1909, the politics of religion has gained momentum. Hindu organizations sought to return anyone of doubtful status to the Hindu fold. Hence, these organizations mobilized tribes to enumerate themselves as Hindus rather than as animists or as followers of tribal religion, as was the case in the earlier census enumeration. A synoptic and insightful view of this politics vis-à-vis tribes is available in Sundar's account of the Indian census (Sundar 1999: 109–110). Notwithstanding the enumeration of tribes as animists, the classification of tribal religion continued until the 1941 census when it was replaced by the category of tribal origin. However, soon after independence, both tribal origin and tribal religion/animism as the basis of census enumeration were done away with. Instead, tribals began to be enumerated as Hindus if they were not adherents of any other major religions. Indeed, one may say that this is the post-independence method of tribal absorption. In the earlier method of tribal absorption, whether described by Bose (1941), Kosambi (1975), Sinha (1962, 1987) or even Srinivas (1977), the agency remained with the tribes, and to that extent they had the freedom to be or not be absorbed. Bose (1941) described the method of tribal absorption as the Hindu method. The Hindu method continues in the post-independence period, but with a difference. That is, in the earlier method, the absorption took place by means of economic co-operation and security, and to that extent the absorption was voluntary. However, in the post-independence period, the absorption has been through the state's administrative practices, which gives no choice to the tribals in declaring their religion (tribal religion) in terms other than those of the major religions. An element of coercion has thus been built into the absorption process.

Articulation of Difference and Identity

The dominant line of thinking in social science inquiries recognizes difference as the hallmark of a distinct tribal identity. However, interaction with non-tribes poses the problem of coming to terms with difference. The interaction, as discussed earlier, eventually leads to the absorption of tribes into the larger society. At the administrative level, despite the constitutional provision for the protection and promotion of tribal languages and cultures, the unstated administrative practices both at the federal and the state levels are geared to securing their absorption into the larger society. This has been most evident in practices pertaining to tribal languages and religions. The domain of languages lies under the jurisdiction of the states and hence forms part of the administrative practices of the states. The domain of religion lies under the jurisdiction of the federal administrative structure and practice. The decision to enumerate tribes as Hindus while abandoning the earlier practice of enumerating them under the 'tribal religion' category was taken at the federal administrative level. Paradoxically, the decision coincided with the adoption of a constitution that provides safeguards for tribes in regard to their language, culture, and tradition. The adoption of these administrative practices at the federal and state levels was the result of the compulsions of regional and national politics. Regional politics aimed at consolidating linguistic and cultural identity by absorbing tribes into it. National politics aimed at consolidating Hindu national political identity, which has been articulated aggressively and coercively since the 1930s.

Despite the attempts at absorption through administrative practices at both the central and state levels over the last 50 years, the attainment of constitutional objectives is still an elusive goal, and is going to become even more elusive in the future. This is evident from the kinds of assertions being made by tribes in different parts of the country today. Of course, such assertions are not uniform. They vary in extent as well as in degree. Whereas they are pronounced in some regions and among some tribes, they are markedly absent in others. However, it is noteworthy that such assertions are on the rise and are at the forefront in the political arena. They can be seen at different levels, but are most clearly discernible in those of land and forest, language and culture, and identity and autonomy. In a way, the assertions of different kinds and at many levels are the result of the increasing social consciousness of difference and identity among tribes. Interestingly, this staking out of identity is taking shape at a time when tribal society is becoming increasingly socially differentiated.

The differentiation among tribes is visible in all areas. For example, one can see it in respect of landholding, which was not the case earlier. Most tribal societies have now become differentiated into the landless, agricultural labourers with some land, marginal farmers, small farmers, middle farmers, and even rich farmers in a limited sense (Shah 1992; Pathy 1984). In terms of occupation, they are now differentiated into cultivators, agricultural labourers, industrial workers, white-collar or salaried workers, and even shopkeepers, traders, and transporters among some tribes like the Khasis, the Jaintias, and the Mizos. Tribes have also become differentiated in terms of access to education, income or wealth and political power. Even in the areas of religion and culture, one can see much differentiation among tribes today. This differentiation is the result not so much of forces from within as it is the outcome of forces from the outside. For example, agrarian differentiation, namely differentiation in regard to access to land, is more the result of land alienation arising from fraud, deceit, indebtedness, and state-sponsored projects leading to large-scale displacement of tribes from their lands and resources (Fernandes and Paranjpye 1997; Singh 2002). Again, occupational differentiation is largely the result of the policies and practices of the state, especially reservation in the areas of education, government employment, and political representation (Xaxa 2001). However, the working of such differentiation is still limited. At the same time, even though it may be limited, social differentiation has caused considerable fragmentation in tribal society. The social solidarity that once marked tribal society has come under stress and strain because of increasing differentiation along the axes of religion, politics, economics, and society.

Despite the fragmentation of tribal society because of social differentiation, the assertion of tribal identity is on the rise. This is largely the result of the emergence of a middle class within tribal society. With the emergence of a middle class, the issue of its culture, tradition, livelihood, control over land and resources, and demand for a share in the benefits of the projects of modernity has become an integral part of identity articulation among the tribes. There is a new kind of consciousness among the tribes now, which has largely to do with the emergence of the middle class. Until recently, the emerging middle class had moved in the direction of acculturation to the larger society through such processes as Sanskritization and Hinduization, which threatened the distinctiveness of tribal society and hence its identity. However, one can now discern a reverse process, which has more to do with alternative avenues of social mobility and social change among tribals. This is mainly because of the opening up of various opportunities resulting from the spread of modern education and the availability of modern occupations (Orans 1965).

The emergence of the middle class is largely the result of the spread of modern education among the tribes. Access to modern education has made it possible for tribal people to gain entry into government and non-government jobs. Tribes have been aided in this venture by the provision of reservation, which is available not only in employment but also in higher education and politics. These measures have made possible the emergence of a middle class in tribal society (Datta-Ray 1983).

The increasing assertion by tribals seen in recent years is the result of cleavages and conflicts between tribes and non-tribes, which have become sharper in the post-independence period. In the early phase of the encounter between tribes and non-tribes, conflict between them invariably assumed the form of revolts, rebellions, and insurrections. Tribal society then was relatively homogeneous and hence had a very strong sense of solidarity. Alien rule was considered illegitimate and hence deserved to be overthrown. However, with the consolidation and expansion of colonial rule in the tribal regions, the winds of change began to blow among the tribals.

As already pointed out, tribes had earlier seen only one kind of change, which was in the direction of absorption into caste society. British rule facilitated movement in this direction because of increased contact between tribes and non-tribes as a result of improved means of communication as well as the processes of urbanization and industrialization. All the same, change in this direction was limited. British rule in the tribal areas also opened up the space for change in another direction. Hence, change in the direction of caste society, and therefore Hinduism, experienced a setback, if not a total halt. This was mainly because Hindus were no longer able to provide protection to tribals, a factor that Bose (1941) considered important. Instead, it led to the domination, subjugation, oppression, and exploitation of tribes. Instead of being cooperative, Hindu practices proved to be competitive and exploitative vis-à-vis the tribals.

There were other factors that arrested this development as well. Of these, the administrative and political concessions extended by the colonial state and the spread of modern education introduced by Christian missionaries, often with the help of the colonial state, were the most influential. While relations with Hindus deteriorated, coming under increasing strain and conflict, relations with Christian missionaries were seen as holding out greater promise. Christian missionaries addressed the problems created by the movement of the Hindu population into tribal areas. It was, therefore, hardly surprising that tribes became increasingly attracted to the Christian faith. By posing the issues of exploitation, oppression, and domination, and addressing the issues of health, disease, education, and language in concrete terms, Christian missionaries heightened the distinctive identity of the tribes as against that of the larger society, especially caste Hindus. It is, therefore, not surprising that Christian missionaries have been accused of depriving the aborigines of their Hindu heritage and of obstructing the natural inclination of the tribes towards Hinduism.

The problems faced by tribes during the colonial period—in the form of loss of control over land and other resources because of alienation of land from tribes to non-tribes and displacement caused by state-sponsored development projects—have continued unabated in the post-independence period. Tribes have been confronted with other problems as well. While post-independence India adopted well-meaning constitutional and legal provisions for their protection, welfare, and development, actual practices have belied this hope. Tribal people have been experiencing deprivation, oppression, exploitation, and discrimination in almost all aspects of life on a daily basis. The government response to the problems of tribes has been lukewarm, indifferent, and often hostile. This experience has been felt most acutely by the emerging middle class. That explains why there has been increasing identity assertion among tribes that now have a sizeable middle class.

The new social mobilization has been prompted by the discriminatory treatment meted out to tribes by the larger population or by the state. Hence, there has been an increasing assertion for securing rights and control over land and other resources in the form of mobilization against alienation of land from tribes to non-tribes and against state-sponsored development projects leading to displacement. What we have been witnessing is the demand for greater political power and autonomy, which has, more often than not, moved in the direction of a demand for a separate state, either within the Indian Union or outside it. This articulation of identity has not been confined to movements for autonomy. It is also seen in the demand for the promotion and revitalization of tribal languages, including the creation of primers and other literature and for their introduction in schools. Related to this is the demand for the development of a script, as in the case of the Santhals and the Tripuri-speaking tribes of Tripura. The choosing of a script from among those with which the tribal people are familiar also forms a part of identity articulation, as in the case of the Bodos of Assam.[6] Identity articulation for greater political power has more to do with the drawing of distinctions between tribes and non-tribes with a view to gaining economic and political power, however limited. The movement connected with language and culture, on the other hand, is primarily concerned with enriching the content of the identity created in the process of interaction between tribes and non-tribes.

The articulation of identity is most pronounced among tribes where an educated middle class has emerged. The larger the educated middle class, the greater is the identity articulation among them. However, the consciousness evident in such identity articulation is not the consciousness of tribe as a category but consciousness of being a people different from others, and especially of being distinct from the dominant regional community. The identity manifested in this articulation is one of being either a Santhal or a Khasi or a Bodo, people generally described as tribes. Often the articulation of such an identity has taken on a more generic name, such as Naga, Kuki, and adivasi. The articulation of the former has been most evident in the domain of language and culture. The articulation of the latter has been most pronounced in the domain of politics.

Conclusions

Despite the constitutional provisions aimed at securing the development of the tribes without causing violence to their languages and cultures, the reality has been quite the contrary. There have been attempts at an aggressive incorporation of tribes into the language and religion of the dominant regional community. Yet, in terms of access to the fruits of development, such incorporation has been, on the whole, avoided and even resisted. The rights of tribes over land, forests, and other resources have been usurped without their receiving any tangible benefits in return. This constitutes the structural setting for the emergence of identity politics among tribes in India.

Identity politics among tribes has generally been described as the politics of tribal identity. In my view, such a description is misleading. Apparently, there is a rupture between the way in which tribal social consciousness is represented by the tribal people themselves and the way in which it is represented by scholars, administrators, and others. The latter invariably regard tribal identity as a means of describing these people. However, such an articulation or expression of identity is not coterminous with the consciousness of tribals. Tribals often have no idea of the category of scheduled tribe or its Hindi or regional-language counterparts. Such an articulation is a part of legal and administrative practices, and constitutes only a small part of the general consciousness of tribals. The articulation of a tribal identity is related primarily to the state and its resources or to the benefits that it may make available. Hence, tribal identity emerges more in the context of the relations of tribal peoples with the state, the facility of reservation, and other forms of affirmative action. In the domain of social and cultural life, tribal identity does not enter as part of tribal consciousness. Hence, tribal consciousness is more a middle-class consciousness than the consciousness of tribals at large.

Notes

1. During the attack on Christians in the Dang district of Gujarat in November–December 1998, the activists of the Sangh Parivar made this point forcefully on the television programmes telecast on the subject.
2. This point too was made during discussions on the television programmes.
3. The similarities have been shown and discussed by colonial administrators in the census reports. Ghurye quotes them profusely in his work on tribes.
4. The terms used to describe tribal religion in India are the same as those used in the context of tribal religion in other parts of the world. The classical anthropological works point to this.
5. Most scholars who have talked of the transformation of tribes into castes have done so more on the basis of practices and rituals than on the basis of their world view or belief system.
6. The script developed and used by the Santhals is known as Ol Chiki. The script developed and used by the Tripuri-speaking tribes of Tripura is known as Kak-Barok. The Bodos are increasingly demanding the use of the Roman script in place of Assamese to mark their distinct identity.

References

Béteille, André . 1960. 'The Definition of Tribe.' *Seminar* 14.

Bose, N. K. 1941. 'The Hindu Method of Tribal Absorption', *Science and Culture* 7: 188–94.

Datta-Ray, B. (ed.). 1983. *The Emergence and Role of Middle Class in North-East India.* Delhi: Uppal Publishing House.

Dumont, Louis. 1998. *Homo Hierarchicus.* New Delhi: Oxford University Press.

Elwin, Verrier. 1960. *A Philosophy for NEFA.* Shillong.

Fernandes, Walter and Vijay Paranjpye (eds.). 1997. *Rehabilitation Policy and Law in India: A Right to Livelihood.* New Delhi: Indian Social Institute.

Ghurye, G. S. 1963. *The Scheduled Tribes.* Bombay: Popular Prakashan.

Kosambi, D. D. 1975. *The Culture and Civilization of Ancient India in Historical Outline.* New Delhi: Vikas Publishing House.

Niyogi, M. B. 1956. *Report of the Christian Missionary Activities Enquiry Committee. Madhya Pradesh, Volume 1.* Nagpur: Government Printing, Madhya Pradesh.

Orans, Martin. 1965. *The Santal: A Tribe in Search of a Great Tradition.* Detroit: Wayne State University Press.

Pathy, Jaganath. 1984. *Tribal Peasantry: Dynamics of Development.* New Delhi: Inter-India Publications.

Ray, Niharranjan. 1972. 'Introductory Address', in *Tribal Situation in India*, edited by K. S. Singh. Shimla: Indian Institute of Advanced Study, pp. 3–24.

Shah, Ghanshyam. 1992. 'Tribal Identity and Class Differentiation: The Chaudhuri Tribe', in *Social Stratification*, edited by Dipankar Gupta. New Delhi: Oxford University Press.

Singh, Chhatrapati. 1986. *Common Property and Common Poverty: India's Forests, Forest Dwellers, and the Law.* New Delhi: Oxford University Press.

Singh, K. Suresh. 1993. 'The Problem in Marginalized Tribals.' *Seminar* 412: 1–7.

———. 2002. *Birsa Munda and His Movement, 1872–1901: A Study of a Millenarian Movement in Chotanagpur.* Calcutta: Seagull Books.

Sinha, Surajit. 1962. 'State Formation and Rajput Myth in Tribal Central India', *Man in India*, 42 (1): 35–80.

———. 1987. *Tribal Polities and State Systems in Pre-colonial Eastern and North Eastern India.* Calcutta: Published for Centre for Studies in Social Sciences, Calcutta by K. P. Bagchi & Co.

Srinivas, M. N. 1977. *Social Change in Modern India.* New Delhi: Orient Longman (reprint edition).

Sundar, Nandini. 1999. 'The Indian Census: Identity and Inequality', in Ramachandra Guha and Jonathan P. Parry (eds.). *Institutions and Inequalities: Essays in Honour of André Béteille.* New Delhi: Oxford University Press, pp. 109–110.

Xaxa, Virginius. 2001. 'Protective Discrimination: Why Scheduled Tribes Lag Behind Scheduled Castes', *Economic and Political Weekly*, 36 (29): 2765–72.

Chapter 8

Protective Discrimination: Why the Scheduled Tribes Lag Behind the Scheduled Castes

The groups identified as scheduled castes and scheduled tribes suffer from certain disabilities. Hence, the Constitution of India made certain provisions for their welfare and upliftment. However, these provisions are far from uniform for the two groups. Indeed, there are more provisions for the scheduled tribes than for the scheduled castes. Articles 15(4), 16(4), 19(5), 23, 46, 330, 332, 334, 335, and 338 are common to the two categories. Articles 29, 164, 244, 244(A), 275(1), 339(1), and 339(2) pertain only to the scheduled tribes. Articles 371(A), 371(B), and 371(C) are in force only in certain states of the north-eastern region, either in the whole state or in a part of it (Verma 1990).

Of the provisions provided for the social and educational advancement of the scheduled castes and the scheduled tribes, the policy of protective discrimination has been one of the most debated and discussed. Based on this policy, the government has laid down three kinds of arrangements. First, seats have been reserved for them in Parliament and the state legislatures. Second, jobs have been reserved for them in government and semi-government services. Third, seats have been reserved for them in educational institutions, especially in institutions of higher learning, such as colleges and universities. To this end, a certain percentage of seats or a quota was earmarked in Parliament and the state legislatures, government services, and educational institutions for members of the scheduled castes and the scheduled tribes. The quota was fixed keeping in mind the size of the population of the two groups in the 1950s when India adopted the policy of protective discrimination. It was 15 per cent for the scheduled castes and 7.5 per cent for the scheduled tribes. The percentage so fixed has remained the same until today, although the population of the two groups has increased. As per the 1991 Census, the scheduled castes constitute 16.5 per cent of the population and the scheduled tribes constitute 8.01 per cent.

Though the reservation policy has been in force for five decades and the population of the two communities has grown over the years, the filling of the quota as fixed by the government is a problem even today. This is far more the case at the upper levels of the government and semi-government services. There are many reasons as to why the quota reserved for the two groups could not be filled. Much has already been written on the subject. The reasons put forward range from apathy and indifference on the part of the government to the unavailability of suitable candidates from the scheduled caste and the scheduled tribe categories on account of their social, economic, and educational backgrounds (Ram 1981). Hence, the focus of this chapter is not on this aspect of the reservation policy. Instead, the chapter will consider the following issues: (1) a comparison between the scheduled tribes and the scheduled castes in relation to the benefits derived by them from the facilities extended to them so far; and (2) an analysis, although tentative, as to why one category has fared better than the other, or alternately, as to why one has not fared as well as the other. It is interesting to note that though the filling of the quota falls short of the government requirement with respect to both the scheduled castes and the scheduled tribes, and particularly with respect to the higher grade services, the availing of the benefits by the two categories has been uneven.

Relative Positions of the Scheduled Castes and the Scheduled Tribes

As pointed out earlier, the provision of reservation for the scheduled castes and the scheduled tribes has been made in three areas: Parliament and the state legislatures (politics), government services, and educational institutions. In politics, the provision of reservation was initially meant for a period of ten years, but ever since then, it has been renewed on a regular basis. It is worth noting here that such time limits, contrary to general belief, were not fixed in the areas of government services and educational institutions (Galanter 1984: 46). Unlike the scheduled tribes, which have been unable to create a space for themselves at the national and regional levels, the scheduled castes have been able to do so, especially in cases where the two groups have been sharing these facilities.

Politics

Let us start with the sphere of politics. As noted earlier, 15 per cent and 7.5 per cent of seats in Parliament are reserved for the scheduled castes and the scheduled tribes respectively. In the state legislatures, the percentage of seats earmarked for them varies according to the size of their respective populations in the concerned state. Now the question is: how have the two categories fared at the regional and national levels? In the areas of government services and educational institutions, the relative positions of the two groups can be discerned from statistical figures. After all, statistical figures are to a greater or lesser extent related to performance. In the area of politics, one cannot judge the relative positions of the two groups in terms of their share in parliament and the state legislatures for two reasons. First, the share of each group is invariably related to the size of its population. Second, and more importantly, there exists in the political sphere a correspondence between what is stipulated and what the actual situation is. In 1999, the number of seats occupied by scheduled caste and scheduled tribe candidates in Parliament was 79 (14.5 per cent) and 41 (7.5 per cent) respectively. The corresponding figures for the state legislatures were 562 (13.8 per cent) for the scheduled castes and 539 (13.2 per cent) for the scheduled tribes (Planning Commission 2000a; Planning Commission 2000b). The higher share of the scheduled tribes in the state assemblies is because of the creation of tribal states, mainly in the north-eastern region. The reasons for the different performances of the two groups in politics, therefore, have to be found elsewhere. Over the question of what these parameters should be, there is surely going to be a difference of opinion among scholars. All the same criteria, such as political appointments to Parliament and the state legislatures or to government posts (such as membership of the council of ministers or the standing committees, or even offices in political parties), may be considered of critical significance in this context. Yet the share of the two categories in the above position may depend greatly on the nature of their participation and the mobilization process. The two together may, therefore, be taken as the domain where the relative performance of the two categories can be assessed. There are, however, no data at our disposal on the aspects corresponding to the ones that we have on government services and educational institutions. All the same, some evidence, although sketchy, is available. In the area of democratic politics, the numerical strength of the groups or categories is of crucial importance, both at the national and regional levels. The numerical strength of the scheduled castes gives them an advantage over the scheduled tribes in national politics. It is true that the scheduled castes and the scheduled tribes do not constitute a homogeneous and cohesive social group either in Parliament or in the state legislatures. They are fragmented along political, linguistic, and regional lines. Yet no government can ignore the issue of adequate representation of groups described as scheduled castes and scheduled tribes. In all these spheres, the scheduled castes have had an edge over the scheduled tribes over the years. The scheduled castes have invariably had better representation in the government in terms of holding ministerial portfolios and other positions like state governorships. Further, whereas the scheduled castes can boast of having produced leaders and administrators such as B. R. Ambedkar and Jagjivan Ram in the past, and K. R. Narayanan, Buta Singh, Ram Vilas Paswan, Kanshi Ram, and Mayawati in more recent times, it is hard to find any leaders of a similar stature from among the scheduled tribes. P. A. Sangma may be considered one such leader, though he does not have a following even within a region, let alone among tribes across regions. In short, the scheduled tribes have not been able to carve out a space akin to that of the scheduled castes in national politics. Even at the regional level, where the scheduled tribes have been sharing power with the scheduled castes, their presence is hardly noticeable.

Government Services

If one examines the relative positions of the two categories in government services, one finds that the scheduled tribes have not fared as well as the scheduled castes in availing of the facility of reservation extended to them by the government. This is shown clearly in the table given below. The data are based on information available as on 1 January 1995.

It is evident from the table that the share of the scheduled tribes in respect of the A, B, C, and D groups of government services was 2.89, 2.68, 5.69, and 6.48 per cent respectively as

Table 6.1: Share of the Scheduled Castes and the Scheduled Tribes at Various Levels of Government Services

Group	Total no. of employees	No. of SC	%	No. of ST	%
'A'	65,408	6,637	10.15	1,891	2.89
'B'	1,08,857	13,797	12.67	2,913	2.68
'C'	23,41,863	3,78,179	16.15	1,33,179	5.69
'D'	10,41,082	2,21,380	21.26	67,453	6.48
Total	35,57,210	6,19,986	17.43	2,05,436	5.78

Note: 'D' excludes sweepers.

Source: Report of the National Commission for the Scheduled Castes and Scheduled Tribes, Vol. 1, 1996–97 and 1997–98.

against 10.15, 12.67, 16.15, and 21.26 per cent respectively in the case of the scheduled castes. The scheduled tribes thus fell short of the stipulated quota both in respect of the aggregate and the individual groups of government services. Moreover, the shortfalls at the higher levels were far greater compared to those for the scheduled castes. In contrast, the scheduled castes fell short of the stipulated quota only in respect of group A and B services, and that too not to a substantial degree. In respect of C and D services, their share exceeded the required quota of 15.5 per cent.

The position of the scheduled tribes with respect to their share in academic or teaching posts in colleges and universities is no better. In Delhi University, until 1995, out of 700 teachers in the university departments, there were only nine teachers from the scheduled caste and the scheduled tribe categories. Out of these, only two were from the scheduled tribe category. If one were to include the undergraduate colleges of Delhi University, the scenario would be even worse. The picture is no different in the other universities in the country, as shown by the data given below.

Table 6.2: Designation-wise Position of SC/ST Teachers in University Teaching Departments/University Colleges, 1992–93

Designation	Total	SC	Percentage	ST	Percentage
Professor	3,338	32	0.96	11	0.33
Reader	5,837	104	1.78	31	0.53
Lecturer	10,191	328	3.22	81	0.79

Source: UGC, 'Representation of Scheduled Castes and Scheduled Tribes in Enrollment, Teaching and Non-Teaching Staff in University Teaching Departments/University Colleges', p. 55.

They are also more sparsely distributed in the field of science and technology compared to other areas of employment. It has not been possible to furnish detailed and comprehensive data in support of the above statement. There is no dearth of data, fragmented though they may be, in support of the point made above. In public sector undertakings such as Hindustan Aeronautics Ltd., Bharat Electronics Ltd., and Bharat Dynamics Ltd., the share of the scheduled castes in Class I jobs was 3.39, 2.38, and 7.90 per cent as against 1.88, 14.37, and 23.70 per cent for Class II jobs respectively. In contrast, the share of the scheduled tribes was 0.12, 0.20, and nil respectively for Class I jobs and 0.26, 0.90, and again nil in respect of Class II jobs respectively (Report of the Commissioner for scheduled castes and scheduled tribes 1975–77).

Educational Institutions

One of the major prerequisites for filling positions reserved for the scheduled castes and scheduled tribes in government and semi-government services is the attainment of the minimum educational qualification specified for these jobs or positions. Hence, reservation has also been provided for in educational institutions, especially those of higher education. However, the scenario with respect to the enrollment of the scheduled tribes in institutions of higher learning such as colleges and universities is almost similar to the situation in the sphere of government services. The following data substantiate this point.

Table 6.3: Faculty-wise Percentage Position of SC/ST Enrollment, 1992–93

Faculty	SC	ST	SC & ST
Arts	8.0	3.9	11.9
Science	5.8	2.1	7.9
Engg/Tech	8.1	2.1	10.2
Medicine	11.9	4.4	16.3
Agriculture	10.7	5.4	16.1
Law	11.7	2.7	14.4

Note: Enrollment faculty-wise includes levels ranging from undergraduate to research plus diploma/certificate courses.

Source: UGC, 'Representation of Scheduled Castes and Scheduled Tribes in Enrollsment, Teaching and Non-Teaching Staff in University Teaching Departments/University Colleges', p. 5.

It is obvious from the above that the presence of the scheduled tribes at the national level is weaker than that of the scheduled castes. In addition, the share of the scheduled tribes in fields such as science, engineering, technology, and law is much smaller than in other fields. The data that we have from the Indian Institute of Technology, Kanpur for the period between 1969–70 and 1973–74 corroborates the existence of this pattern. Since then there has been marked improvement in the share of the tribes.

Table 6.4: Admission of Scheduled Caste and Scheduled Tribe Students in IIT, Kanpur, 1967–78

Year	No. of candidates applied for JEE			No. of candidates in merit list? and called for interview		No. of candidates offered admission		No. of candidates actually admitted		
	Total	SC	ST	SC	ST	SC	ST	SC	ST	Total
1967–68	62	47	15	–	–	–	–	–	–	–
1968–69	47	2	–	–	–	–	–	–	–	–
1969–70	35	29	6	7	1	7	1	7	1	8
1970–71	28	–	–	6	1	6	1	6	1	7
1971–72	35	–	–	1	–	1	–	1	–	1
1972–73	79	–	–	1	–	1	–	1	–	1
1973–74	98	86	12	42	6	42	6	–	–	46

Source: K. N. Sharma et al., Educational Arithmetic of Social Inequality: A Study of Admission and Adjustment of IIT, Kanpur Students, 1974 (mimeograph).

One finds that the enrollment of the scheduled tribes in various courses offered at the higher levels of education falls below that of the scheduled castes. Moreover, the share of the scheduled tribes in the various courses offered is very uneven. One finds that in comparison to the natural sciences, more scheduled tribe candidates prefer the arts and humanities. This is apparent from the enrollment data given above as well as from the employment data of the scheduled tribes in sectors, departments, and faculties where science and technology training is crucial. The employment pattern of the scheduled tribes in public sector undertakings has been referred to earlier. The pattern is similar in colleges and university departments. In the North-eastern Hill University, Shillong, for example, of the 30 teaching positions held by scheduled tribe candidates in 1986, only three were in the faculty of sciences. The scenario at Delhi University or elsewhere is no different. Even within the sciences, more scheduled tribe candidates opt for courses such as agriculture and medicine rather than other science courses. Of equal significance is the fact that the enrollment of scheduled tribe members in arts and humanities courses, contrary to general belief, is lower than it is in medicine and agriculture courses. The situation in NEHU since then has improved. There are about 20–25 scheduled tribe teachers in the faculty of sciences in NEHU today. Even at the regional level, where the scheduled castes and the scheduled tribes share in resources, the performance of the scheduled castes, as can be seen in the table below, has been invariably better. Paradoxically, this is the case even in those states where the scheduled tribes easily outnumber the scheduled castes. Madhya Pradesh and Orissa may be taken as two appropriate cases. The scheduled tribes constitute between 22 to 23 per cent of the total population in each of the two states as against 4 per cent and 7 per cent for the scheduled castes respectively.

Table 6.5: State-wise enrollment of SCs and STs in UG and PG Courses

State	Undergraduate		Postgraduate	
	SC	ST	SC	ST
Gujarat	3.6	1.9	6.4	12.8
Karnataka	23.3	2.6	21.7	6.4
Madhya Pradesh	13.9	6.1	12.5	4.8
Maharashtra	11.4	3.0	12.1	2.0
Orissa	14.1	4.8	13.1	5.2
West Bengal	11.2	0.5	7.3	0.4

Source: UGC, 'Representation of Scheduled Castes and Scheduled Tribes in Enrollment, Teaching and Non-Teaching Staff in University Teaching Departments/University Colleges', pp. 12–18.

Explaining the Phenomenon

Though reservations have been provided for the scheduled castes and the scheduled tribes, the filling of these seats, especially in government services and institutions of higher education, is not routine and mechanical. Rather, it is dependent on the candidates' performances in tests, examinations, and interviews, besides, of course, in their meeting the appropriate educational requirements. This partly explains why the state has not been able to fill these posts and seats for both the scheduled tribes and the scheduled castes. This has been the case especially with the scheduled tribes. It is striking that the scheduled tribes have invariably played second fiddle to the scheduled castes when it comes to taking advantage of the benefits and facilities extended to them, whether it is in education, science and technology, the civil services, or politics. An attempt is made here to examine why the scheduled tribes have not achieved as much as the scheduled castes. The aim is not to examine the basic assumptions underlying the educational system, with its overriding focus on individual performance and competitiveness. The provision of reservation by itself does not question the basic premises.

Why have the scheduled tribes not performed as well as the scheduled castes? Does the difference in their relative performances have anything to do with factors such as demographic size, literacy rate, and geographical mobility? It is true that the scheduled caste population is twice the size of the scheduled tribe population, but for our purposes here, this factor is hardly significant, as the issue under examination is in keeping with the size of the respective populations of both groups. In terms of overall literacy, the scheduled castes have fared better than the scheduled tribes. As per the 1991 Census, their literacy rate was 37.4 per cent as against 29.6 per cent for the scheduled tribes. In 2001, the literacy rate of the scheduled castes stood at 54.34 per cent as against 46.42 for the scheduled tribes. The geographical mobility of the scheduled tribes, especially the educated sections, is not as pronounced as that of the scheduled castes both in regard to intra and inter-state migration. within and outside of the states. It is precisely these differences that need to be problematized rather than be taken for granted or ignored as corollaries of the reservation policy.

Any attempt at understanding such problems entails a discussion of two aspects or levels, which are intimately related to one another. The relative positions of the scheduled castes and the scheduled tribes in availing the benefits provided by the government are to a great extent dependent on two factors. One is the relative position of the two groups with respect to their general educational attainment, especially at the level of higher education. The other factor is their overall performance at these levels. For it is not only their enrollment in higher education courses, but also the quality of their performance over the duration of the courses and their successful completion of the same that is critical in assessing their eligibility for jobs and positions at different levels. It may not be out of place to point out here that the dropout rates for the scheduled tribes between classes I and X are higher than those for the scheduled castes. The figures for the scheduled tribes in 1980–81 and 1990–91 were 91.18 and 86 per cent respectively as against 86.91 and 80.58 per cent for the scheduled castes (Planning Commission 2000a; Planning Commission 2000b).

Before one explores the above question, it is important to note that a certain degree of arbitrariness is invariably involved in dividing the population into categories such as scheduled castes and scheduled tribes. There have been groups, small though in number, which have been categorized as tribes not because they share attributes characteristic of tribes but on considerations such as political and administrative manoeuvring. Such arbitrariness extends to the demarcation of the population into various other categories as well. However, they are more exceptions than the general rule.

One could begin exploring the above question by examining the logic of the reservation policy for the two categories. It must be mentioned here that the logic of reservation for the two categories, in contrast to conventional belief, is not identical. The scheduled castes were extended reservation because they suffered from certain disabilities arising from their segregation from the dominant community. The scheduled tribes, on the other hand, were granted reservation not because they were segregated from the dominant community but because they lived in complete isolation from the dominant community. This difference is hardly ever pointed out, let alone explicated. Rather, the general tendency is to treat the two similarly.

This may be taken as the starting point for a discussion as to why the scheduled castes have fared better than the scheduled tribes. The scheduled castes generally have been described as those falling outside the varna system. Though outside the varna system, the scheduled castes, or the former untouchables, are integral to the dominant community, whether in the village or the town, the only difference being that the untouchables were unjustly segregated from the community and were subjected to forms of disabilities and discrimination. The scheduled castes have, therefore, had greater exposure to the larger society as compared to the scheduled tribes. This is to say that the opportunities made available to the larger society or to the higher castes in the form of knowledge, information, technology, employment, etc. were also in sight for the scheduled castes, even though they were denied access to it. Such advantages were absent for the scheduled tribes. The tribes thus suffered the disadvantage of isolation in this regard.

It is highly unlikely that educated Indians are unaware of the categories 'scheduled castes' and 'scheduled tribes'. Yet, while they may be familiar with the scheduled caste category and have an idea of what it means, the same cannot be said, with a similar degree of certainty, about the scheduled tribe category. To begin with, despite being segregated, discriminated against, and oppressed, scheduled castes are seen invariably as a part of Hindu society. Since Hindu society can be understood effectively at the regional level, scheduled castes are an integral part of the regional community. This regional community is also at the same time a linguistic community. Hence, though divided on the basis of caste, scheduled castes share the language, culture, tradition, etc. of the larger linguistic community. They constitute a part of the dominant community of the region. This is not the case with the scheduled tribes, and this turns out to be one of the most serious handicaps before them. Indeed, an average Indian cannot figure out where the scheduled tribes fall in the social structure. Members of tribes are, therefore, invariably seen as outsiders, as those who not only do not speak their language but also as those who do not share their customs, traditions, and values. Belonging to the scheduled tribes, though advantageous at the local level, is hardly so at the regional and national levels. Indeed, tribes suffer from all the disadvantages associated with ethnic minorities and communities. Hence the scheduled castes have generally been able to fill the jobs earmarked for them at the lower levels. Jobs allocated for tribes at such levels, however, remain vacant mainly for two reasons. Either scheduled tribe candidates are generally not available, or when available, face general discrimination or prejudice being outside of the majority or dominant language and religion. Further, to get these jobs, members of the scheduled tribes also have to learn the language of the dominant community, an advantage that the scheduled castes already possess. In short, whereas scheduled castes are acceptable to the dominant community, scheduled tribes are not.

The presence or absence of exposure, though important, is not a sufficient explanation for the differentials between scheduled castes and scheduled tribes in their ability or willingness to make the most of reservations. The scheduled tribes, as we know, is not a homogeneous category. Not only do the tribes differ in terms of size, language, regional origin or religion, they also differ in terms of their exposure to the larger society. The contact of the tribes with the larger society, though not altogether absent, was tenuous and limited prior to the arrival of the British. The situation changed with the onset of colonial rule. Tribes were increasingly brought into contact with the larger society through modern means of transportation and communication like roads and railways, growth of trade and commerce, and expansion of market forces. This process intensified in the post-independence period as a result of the geographic, economic, and political integration of the country. Of course, the degree and nature of this integration varies from tribe to tribe. Yet tribes with a history of such contacts and exposure to the larger society, especially lasting more than a century, have not necessarily fared better than those that have had minimal exposure.

Hence, to seek the reasons as to why scheduled tribes have not fared as well as scheduled castes, we have to go beyond the aforementioned explanations. My contention is that exposure per se is not as important as the social structure they represent and of which they are a part. Besides being considered outside of 'civilization', tribes in India represented a different type of society. Tribal societies on the whole are small in size and are marked by a great deal of homogeneity despite being differentiated along the lines mentioned above. They are characterized more by similarity and resemblance than by differentiation and heterogeneity, especially in terms of their structural features. Thus, tribal societies are marked by an absence of interdependence, division of labour and occupation, and correspondingly by a heterogeneity of values, skills, knowledge, income, wealth, status, and privilege.

The structure specific to tribal society gives rise to certain disadvantages. Tribes are small-scale societies and are not part of the larger society, though they may have some kind of contact and interaction with it. The description of tribal society as small scale is relative, that is, in relation to the dominant community, but it varies in size, from over 7 million in the case of the Gonds and the Bhils to less than 1,000 in the case of many tribal communities. Nevertheless tribal societies are characterized by a great deal of homogeneity in their social structure, a point that has already been made. Interestingly, there seems to be no correlation between the size of the population of the different tribal groups and their performance in literacy and other development related indicators. Rather, the pattern seems to be that the larger the size, the lower is the literacy rate, educational attainment, and share in the facilities extended by the government. This means that the size of the population has nothing to do with the differential share of tribal groups with respect to the facilities extended through reservation. The case of the Gonds, the Bhils, and the Santhals, the three major tribal communities in India, will be examined in this context.

A lack of heterogeneity in terms of the social division of labour, occupation, skill, class, access to power and modes of living has been considered characteristic of tribal societies. This being the case, there is hardly any group or stratum that can be taken as the reference point for emulation by the scheduled tribes. Of course, there have been tribal societies with some forms of stratification, if one may use the term, but they have been rudimentary and mainly based on clan or lineage. Moreover, they were not characterized by status differentiation, though privileges of some kind were not ruled out entirely. Further, there is hardly anything like tribal identity among the scheduled tribes at the pan-India level. If at all there is anything like tribal identity, it is confined either to a locality or region, or at best to an institution or organization. Moreover, such an identity is at work more at the levels of the political or interest articulation than at the social or cultural planes. Even when there has been an identity such as this, it has not been assertive and active. Such an identity among scheduled castes, both at the regional and national levels, is more pronounced in comparison to that of the scheduled tribes. One of the reasons why such an identity is much stronger among the scheduled castes is that they have gone through a common historical experience of oppression and discrimination for centuries. This was not the case with the scheduled tribes. Further, the oppression and exploitation of the tribes took an economic and political form. It did not percolate to the social and cultural levels, as was the case with the scheduled castes. The reason for this is that the scheduled castes were part of the larger community, where as the scheduled tribes were not. The scheduled castes, therefore, have had reference points not only within the system of which they were a part, viz. higher caste groups, but also within their own social category. They have been able to develop a very strong sense of identity both at the regional and national levels despite being regionally and linguistically heterogeneous. This has been helpful and important not only in social networking, but, more importantly, also in the areas of interest articulation and social and political mobilization.

Tribes have now been exposed to the market, the bureaucracy, the state, and the wider world. But earlier they had little knowledge of or interaction with these outside forces, largely because they had little to do with people they considered different. Even if they had to interact with outsiders, they did so minimally. Not only did the non-tribes treat the tribes as outsiders, the tribes themselves considered the non-tribes outsiders, and hence not worthy of emulation. Of course, there were exceptions to this rule. The incorporation of tribes into the larger society paved the way for more intense interaction between tribes and non-tribes, leading to increasing exploitation of and discrimination against tribes. This was, broadly speaking, confined to the economic sphere. In the social and cultural spheres, tribes lived a life outside of the larger society. This is not to deny that certain sections among the tribes sought to adopt the lifestyle of the castes and groups of the larger society. Such emulation, however, remained confined to the cultural sphere and failed to be adopted at the structural level. Where it did succeed in being adopted, tribes by virtue of being absorbed into the larger society lost their distinct identity. The desire for integration with the larger society, and the impulse to be identified with it, was, on the whole, missing. The larger society was seen as exploitative and oppressive, something to be opposed and fought against. In the process, it could never become a standard for evaluation, emulation, and aspiration.

An equally important aspect worth noting is that it is collectivity and not individuality that dominates the larger part of tribal life. Hence, the principle of individuality, and the celebration of individual performance and excellence, is less acceptable as the ethos of tribal societies. Collectivity is still the central value or ethos among tribes despite the existence of certain forms of inequality among them. Of course, with their incorporation into the wider world and the provision of protective discrimination aimed at the advancement and welfare of tribal people, tribal societies are becoming increasingly differentiated. The processes are, however, uneven among them. Despite these developments, egalitarianism to a greater or lesser degree continues to be the guiding ethos or spirit of tribal societies. In view of values such as these, the significance of the individual and of individualism does not assume the place that it assumes among other groups. Hence, performance and achievement stemming from the spirit of individualism are not reflected adequately in tribal societies. That such a tendency or spirit is also partly responsible for the poor academic performance of tribal students cannot be denied.

It is also important to bear in mind that tribes are by and large dependent on nature, viz. land and forest, for their survival and livelihood. They make their living mainly as food gatherers, hunters, livestock holders, cultivators, or follow a combination of these occupations. In view of this, magic and ritual, or at best rational adaptation rather than rational calculation and domination, characterize their orientation towards life. The fact that the scheduled tribes are relatively better represented in terms of enrollment and employment in disciplines connected with agriculture and medicine may have something to do with the link that they have with the organic world. In comparison, the scheduled castes are less rooted in nature and hence are more attracted to abstraction the sine qua non for science and mathematics. The scheduled castes are engaged in occupations that are either service or craft-oriented; these occupations are characterized to some degree by calculation, reciprocity, and commitment, and hence go beyond the rhythms of nature. This may partly explain the higher ratio of enrollment and employment of the scheduled castes in the fields of science and technology in comparison to that of the scheduled tribes. By virtue of being engaged in such fields, they had the advantage of being absorbed in industries that came up during the period of British rule. They thus had the historical advantage of being not only a part of modern industrial organizations but also of participating in the processes of urbanization and urbanization, that were accelerated under British rule.

Given the structural setting in which the scheduled castes lived compared to that of the scheduled tribes, the constitutional provision turned out to be a major breakthrough for them, but not for the scheduled tribes. With the adoption of the Constitution, the ideology of hierarchy and inequality was formally replaced by the ideology of equality and social justice. For the first time in the history of India, the scheduled castes were offered the same civil and political rights as the other citizens. The harmonic system, viz. congruence between the ideology and the social structure that characterized Indian society, gave way to a disharmonic system with the adoption of the Indian Constitution (Béteille 1974). Now there was disharmony between a structure that was inherently non-egalitarian and an ideology that called for the abolition of discrimination based on race, caste, religion, etc. The actual social system, then, was far from conducive to the realization of the ideal that was set in motion in independent India. Hence, various kinds of provisions were extended under the Constitution for the benefit and protection of those affected by the system. Of these provisions, reservation has come to be the most important. As noted above, the scheduled castes were a part of the larger society and had faced discrimination at almost all levels of social life for centuries. The extension of the constitutional provision marked the beginning of the phase aimed at reversing this process. Hence, the scheduled castes aggressively seized the opportunities made available by the Constitution. The fact that they were a part of the dominant community, and were exposed to whatever the dominant community was exposed to, helped them in taking advantage of the facilities extended by the government under the constitutional provision. Their position in the social structure helped the scheduled castes in availing of these benefits. Since discrimination was experienced at all levels of social life, the scheduled castes made use of any opportunity provided to them for reversing such an arrangement. Since their traditional occupations determined their low social status, they were desperate to abandon these. In the context of new and diversified occupations made available under the British, the processes of industrialization and urbanization assumed critical importance, opening up space not only for geographical mobility but also for social mobility in view of the increased scope for interaction provided for the scheduled castes. Indeed, industrialization and urbanization acted as major catalyst. The adoption of the Indian Constitution in 1950 and the constitutional provision for the scheduled castes therein gave a major boost to the process of social and geographical mobility unleashed by the forces of industrialization and urbanization.

This was not the case with the scheduled tribes. They did not experience the kind of discrimination experienced by the scheduled castes, primarily because they were not an integral part of the dominant society. What they did experience under colonial rule was exploitation at the economic level and domination at the political level. They sought to fight this through a stronger articulation of their rights over land and forest, on the one hand, and a demand for political autonomy, on the other hand, rather than by demanding a share in government jobs. The Constitution, however, did not extend such a guarantee to them. Industrialization did not benefit them in the way that it did the scheduled castes, mainly because of the nature of their occupations. The scheduled tribes also did not dislike their occupations as was the case with the scheduled castes. First, there was no stigma attached to their occupations. Second, there was little inclination to abandon traditional occupations (agriculture, hunting, and food gathering) among the tribes. They were more interested in acquiring land and pursuing life as agriculturists at the first opportunity made available to them. Third, tribes face disadvantages that have accumulated with the passage of time. Since they have been independent of the larger community, and hence are not part of the dominant linguistic and religious community, they have invariably been seen as alien to that society, and hence not as legitimate claimants to a share in the state's resources. They live mostly in relatively inaccessible areas, and hence have been the last to be exposed to the modern world, especially the state, along with its legal–bureaucratic structures, and the market. Even when they came to be exposed to these forces, it was not as a part of the social structure but as the victims of it.

In the sphere of politics too, as noted earlier, the impact made by the scheduled castes at the national and regional levels has been greater than that made by the scheduled tribes. This is partly because of their numerical strength, but, more importantly, because of the better articulation of their demands, more effective manoeuvring strategy, and better mobilization. The mobilization process in the case of the scheduled castes, unlike that in the case of the scheduled tribes, cuts across districts, regions, and even states. In the course of time, the latter have invariably emerged as national leaders, with followers from social groups cutting across different regions and states.

Since the scheduled castes have been a part of the dominant community, they are spread over the length and breadth of the country. Hence, their presence is felt much more strongly at the national and regional levels. Their presence becomes even more visible and significant because of their numerical strength. In addition, they enjoy greater unity compared to the scheduled tribes because of certain common experiences that they have undergone, irrespective of the region or the state to which they belong.

Unlike scheduled caste leaders, the impact made by scheduled tribe leaders, even when they are nationally recognized figures such as P. A. Sangma, has been less visible. It is most evident at the level of the articulation and championing of the tribal cause. Tribal leaders, including Sangma, are unable to pursue and protect the interests of the scheduled tribes as aggressively as the scheduled caste leaders. Sangma is not seen as a liberator by the tribes he represents, which is evidently the case with Paswan, Kanshi Ram, and Mayawati in relation to the scheduled castes. The reasons for this are rooted mainly in the nature of tribal society, which within any given region is divided along the lines of language, territory, and culture, on the one hand, and the needs, interests, and aspirations of its various constituent elements, on the other hand. The scheduled tribes also face difficulty in negotiating with members of the larger community, for the latter invariably see them as alien and do not regard them as an integral part of their society. Hence, there has been an inherently centrifugal rather than centripetal tendency among the scheduled tribes in India. At the sub-regional or local levels, the scheduled tribes have, however, succeeded in making an impact to some degree, but only in areas where they are in a numerical majority.

Why Some Tribes have Fared Better than Others

As mentioned above, the scheduled castes have, in general, fared better than the scheduled tribes. But within these categories, some castes or tribes/communities have fared better than the others. Among the scheduled castes, for example, the Mahars of Maharashtra and the Chamars and the Jatavs of north India have done better than the other groups. Similarly, among the scheduled tribes, there are some that have done better in comparison to other tribal groups, if not as well as the scheduled castes. Indeed, there are cases where some tribal groups have done better than some scheduled caste groups. It would be interesting to explore the reasons for such differences in the case of these groups. The focus here is not so much on castes as it is on tribes. Hence, differences among the scheduled castes have not been highlighted as much as the differences among the scheduled tribes. Some tribal communities that have done relatively well are the Minas of Rajasthan, the Mizos of Mizoram, the Khasis of Meghalaya, and the Oraons of Bihar and Madhya Pradesh. Of these, the Minas are particularly noteworthy.

The Minas and the Oraons, with a population ranging between two and three million each, are smaller than groups such as the Gonds, the Bhils, and the Santhals. The Khasis and the Mizos are even smaller groups. What seems to account for the better representation of the Minas in government services is that they constituted a fairly differentiated community. Among them were large and small landowners as well as chowkidars. They were also differentiated into other categories depending on factors such as their geographical spread, nature of their social relations with caste groups such as the Jats, the Ahirs, and especially the Rajputs, and the adoption of specific occupations. As land lords or zamindars, many Minas lived off the land as extractors of rent. As lords, they were also in more regular interaction with their counterparts in other communities and also with the state authorities. The privileged section of the Mina community was thus in a position to take advantage of opportunities that were also available to similarly placed sections in other communities. The provision of reservation gave an added advantage to members of the Mina community. Once such processes were set in motion within the community, the less privileged among them also began to take the path set by the more privileged.

Besides the Minas, other tribal groups that have done relatively well are the Mizos and the Khasis of north-eastern India and the Oraons of eastern and central India. The Mizos and the Khasis with a population of less than one million each were ruled by chieftains. They also observed a degree of social differentiation based on rank, status, and lineage. The spread of Christianity and modern education opened up new opportunities for those seeking higher rank and status. The growth of townships in the tribal heartland resulting from the establishment of new administrative systems and the rise of trade and commerce gave not only a major boost to the process of social differentiation in these societies but also led to intense interaction among the groups living there, including the tribes. There was also a strong correlation between urbanization and a higher literacy rate among the tribes.

The Oraons, unlike the Mizos and the Khasis, did not have a strong system of chieftains and kings, and hence was relatively egalitarian. The spread of Christianity and modern education among them, however, opened up the path towards social differentiation, viz. the distinction between Christian and non-Christian, educated and non-educated. It also introduced a certain degree of occupational differentiation within the community, seen in the emergence of a priestly class, teachers, and other occupational groups, which were accorded a high status. The emergence of such groups fairly early helped communities such as the Oraons to take advantage of the services provided by the modernization process. This explains, although only partially, why the Oraons have done better than the others in availing of the facilities provided by the government under various constitutional provisions.

These then are some of the structural features that help explain the better performance of the scheduled tribes and the scheduled castes.

References

Béteille, Andre. 1974. *Studies in Agrarian Social Structure*. New Delhi: Oxford University Press.

Galanter, M. 1984. *Competing Equalities: Law and the Backward Classes in India*. New Delhi: Oxford University Press.

National Commission for the Scheduled Castes and Scheduled Tribes. 1998. Report of the National Commission for the Scheduled Castes and Scheduled Tribes, Vol. 1, 1996–97 and 1997–98.

Planning Commission. 2000a. *Tribes in India (A Data Sheet)*.

———. 2000b. *Scheduled Castes in India (A Data Sheet)*.

Ram, Nandu. December 1981. 'The Limited Education.' *Seminar* 268: 20–24.

Report of the Commissioner for Scheduled Castes and Scheduled Tribes, Part II, 1975–76 and 1976–77 (Appendices).

Sharma, K. N. et al. 1974. 'Educational Arithmetic of Social Inequality (A Study of Admission and Adjustment of IIT Kanpur Students)', Department of Humanities and Social Sciences, Indian Institute of Technology, Kanpur (mimeograph).

University Grants Commission. 1992–93. 'Representation of Scheduled Castes and Scheduled Tribes in Enrolment, Teaching and Non-Teaching Staff in University Departments/University Colleges' (mimeograph).

Verma, R. C. 1990. *Indian Tribes Through the Ages*. New Delhi: Ministry of Information and Broadcasting, Government of India.

Chapter 9

Tribal Culture and Ecology: The Changing Dimensions

Ecology is a study of the relationship between organisms and their environment. The concern in sociology with ecology has to do with the relationship between human beings and their environment. Human beings live in groups distinct from one another in respect of their social structure and culture. Hence, the focus in sociology is invariably on the relationship between groups or communities and their natural environments. This relationship is twofold. The culture of a group or community is shaped by the environment. At the same time, human beings too have an impact on the environment. The relationship between the two is far from uniform. There are countries where the relationship is balanced and others where it is exploitative. The same is the case in respect of regions and groups or communities within a given country. India is no exception to this.

Rapid advances have been made at the cost of natural resources. The greater the development, the greater the destruction of natural resources. Increasing realization of the danger of unchecked ecological and environmental damage has prompted a renewed interest in the conservation of the environment, and even its regeneration. The development of environment-friendly technology requires enormous resources, which only the rich and industrialized nations possess. Further, such technology is still in the nascent stage of development. At the present stage, the major thrust of the development agenda is the preservation of the natural environment. Hence, techniques based on traditional knowledge systems are being sought. That partly explains the renewed interest in the study of tribal people. After all, tribal people are seen as living in relative harmony with the natural environment. The interest in tribal people is thus not so much as people but insofar as they contribute to the maintenance of the natural order. The social and cultural practices by which they relate to the environment and preserve natural resources have come to be one of the major interests in environmental and ecological study (Hebbar 1995).This is abetted by the fact that social and cultural practices and values associated with developed societies have come to be linked with comfort and luxury, which people in these societies are not ready to sacrifice even though they recognize the importance of maintaining the natural environment. Rather, more often than not, they tend to pursue an agenda of exploiting the natural environment, the cost of which they are not interested in bearing but which they want other people, namely the tribals, to bear. This is the case in India, as the discussion below will show.

The way in which tribes have been conceptualized has a great deal to do a with their physical, social, and cultural proximity with nature. Tribes have thus been defined as those who live in simple or primitive conditions, primarily dependent on nature for their survival and livelihood. They have been seen as having a symbiotic relationship with nature. Even deserts, where pastoral tribes lived, were seen as having formed cul-de-sacs, ensuring thereby the physical and cultural survival of the communities (Roy Burman 1993: 180). Tribals are seen to have traditionally balanced their needs with ecological imperatives. They preserved forests as a resource for posterity, it has been argued. They had what Fernandes calls a constructive dependence on forests and other natural resources. However, as a result of the vicious cycle initiated by industrial clear felling and displacement for development projects, there has been a transition to destructive dependence on natural resources (Fernandes 1993: 48).

The relationship of tribes with nature can be explored in two ways. The first is the way in which nature and the environment are articulated and represented in riddles, stories, myths, legends, feasts, and festivals. The focus here is on the way in which the environment is articulated and represented in the cultural domain. The second is examining the ways in which the tribes relate with nature and the environment in their day-to-day existence, that is, the existential conditions of their living. These dimensions are explored in the following discussion.

Tribes in India, though at different stages of development, still live primarily off the land and forests. Some tribes are still practice food-gathering and hunting, and depend exclusively on the forest for their survival. In administrative parlance and practice, they are defined as 'primitive' tribes. Since the Fifth Five Year Plan period, the government has identified 75 such tribal groups in India. The number continues to be the same even after more than three decades of special development projects meant for them. The majority of tribal groups are agriculturists. They practice two kinds of agriculture, settled agriculture and shifting cultivation. Shifting cultivation is most widespread in north-eastern India. In other parts of the country, some tribes inhabiting the areas adjoining the states of Orissa, Andhra Pradesh, and Chhattisgarh practice shifting cultivation. Needless to say, shifting cultivation is linked with forest clearance. The more prevalent form of agriculture among tribes is the same one practiced by non-tribal populations, namely settled agriculture. The only difference is that tribals even then depend on forests for a large part of their needs. Hence, despite being settled agriculturists, their dependence on forests is considerable, which is not the case with non-tribal agriculturists.

Existential Dependence

The tribals' dependence on nature is evident from their daily livelihood needs, beginning with food. Their food habits are intimately linked with their relationship to nature. The staple diet of settled agriculturist tribes consists of rice, dal, and vegetables. Yet what is striking is that leaves, flowers, seeds, roots, and fruit form an integral part of their daily diet. These are procured from the forest. Only a few are cultivated by the tribes themselves. A study has shown that the Oraon tribes eat the leaves of about 21 common native plants. The number of common native plants consumed by the Oraons either whole or in part (flowers, roots, seeds, and fruit) is considerable. In all, 87 common native plants feature in the diet of the Oraon tribes (Kujur 1989). Hoffman (1950) enumerates 71 different wild plants used by the Mundas as pot-herbs, 26 of whose tubers, fruit, and roots are used as vegetables, 15 trees and shrubs whose young leaves are used as pot-herbs, 10 others whose leaves are eaten raw, and 25 wild trees and plants whose flowers are used as vegetables. He also lists 17 cultivated tubers and roots that are used as food, 28 plants cultivated for other purposes and used also as pot-herbs, and 14 plants cultivated as pot-herbs (Fernandes 1993: 49). In regions where shifting cultivation is practiced, food-value trees and plants along with shifting cultivation constitute more than 50 per cent of the food. It is estimated that at least 50 per cent of the food requirements of tribes are satisfied by produce from forests; some scholars place the figure as high as 80 per cent (Fernandes 1993: 48).

Besides food, tribals are also dependent on forests for construction material for their huts or houses and for raw material for tools and implements required for daily economic and household activities. In the case of tribes dependent on hunting, fishing, and food gathering, shelters are generally of a very elementary nature; poles, frames, and roofing materials are all obtained from the forest. The houses of agriculturist tribes generally have mud walls and tiled or thatched roofs, but even among them there is considerable use of bamboo and timber as poles and frames. They are also dependent on the forest for their daily tools and implements. Agricultural implements like ploughs and yokes, devices used for lifting water for irrigation, and threshing and winnowing tools are all made from materials obtained from the forest. That is also the case with tools and weapons required for hunting, fishing, and food gathering. Hunting implements such as bows and arrows, slings, and spears are made from forest produce. Similarly, fishing tools such as baskets and traps of various kinds are made of bamboo. Fishing nets are made of twines. Umbrellas are made with the handles and ribs of bamboo covered with leaves. Hooded waterproof coats are made of leaves (*gungu*). Most tribal households have items such as paddy-husking contraptions, mats, cots, wooden stools, baskets, cups, plates, cushions, ropes, mortar and pestles, and oil presses, all of which are made from forest produce. Even art objects, artifacts, musical instruments, and ornaments are made from forest produce. In fact, the whole material culture of tribes is rooted in the tribal people's dependence on forests and forest produce. Their ties with forests cannot be severed without destroying their entire way of life.

Knowledge of the treatment of diseases is another sphere where we find a close relation between the tribal communities and their environment. Treatment of diseases is invariably based on the use of medicinal herbs found in forests and adjoining areas. The Oraons use such medicines to treat about 34 diseases, including remedies for pain (headache, toothache, stomachache, eye pain, ear pain, and migraine), fever (high, ordinary, malaria), wounds, constipation, diarrhoea, dysentery, epilepsy, rheumatism, insomnia, tetanus, and eczema. These are treated with medicines derived from the leaves, roots, and barks of plants and trees that grow wild in the jungle. Some of these are cultivated in fields by tribal people (Kujur 1989: 138–44). In a study on Orissa and Madhya Pradesh, Fernandes and his colleagues identified 40 trees and more than 50 herbs used for medicinal purposes in these areas. A study of tribal medicine in Kerala identified at least 39 species of roots, 15 of fruit, 30 of leaves, 12 of bark, and many kinds of latex, flowers, and other plants and herbs in common use (Fernandes 1993: 51).

In addition to being the very source of their identity and way of life, tribals and other forest dwellers derive a large portion of their monetary income and livelihood from forests. One of these sources of income is firewood. It is no wonder that state and forest department officials have accused tribes of destroying forests. However, this is a recent development. The main source of income for the large majority of tribals is not firewood but what is described in administrative records as minor forest produce (MFP). Scholars and activists, however, criticize the use of such a description, arguing that for tribals this is not a minor but a major source of income. Hence, they prefer the term non-wood forest produce (NWFP). Some of the major NWFP includes lac, kath, bamboo, and leaves such as tendu, keond, sal, and gungu. The mahua flower and fruit are also important sources of income for tribals. Needless to say, non-timber produce has different values in different places and regions. The collection, processing, and sale of NWFP and firewood provide several days of employment for tribals (Fernandes 1993).

Another mode of dependence on forests is slash and burn or shifting cultivation, which is widely practiced in the hill regions of north-eastern India. The practice is more popularly known as jhoom. This agricultural or livelihood practice is also followed by some tribes in certain pockets of eastern and central India. Jhoom has come under sharp criticism from development planners and environmentalists, who argue that the practice of shifting cultivation destroys forest wealth by causing deforestation. They suggest that tribes should be advised to switch over to either terraced cultivation or horticulture. The arguments and suggestions are made, however, without giving adequate and careful attention to the problem. As Misra says, if deforestation is the argument, it has to be shown that more forests are needed, that forest-growing is more profitable than crop-growing, that growing forests is cheaper, and that a market for forests exists. Alternatively, it has to be shown that cultivable land either exists or is available elsewhere or that gainful employment for forest dwellers in other occupations is feasible.

Another related criticism is that shifting cultivation leads to soil erosion. However, even this argument, as Misra puts it, is made without examining whether the rate of soil formation is less than that of soil erosion, or if soil erosion stops without jhooming, or, alternatively, if horticulture and terrace agriculture prevent erosion, land is fit for terraced cultivation and jhooming, however long the jhoom cycle may be, (Misra 1976: 80–84). Those who subscribe to the view that shifting cultivation results in soil erosion will find it disquieting to learn that shifting cultivation is, in fact, a response of the tribal people of the hill areas to the problem of erosion of fertile soil located on steep slopes. In this regard, M.S. Sivaraman, Adviser to the Programme Administration of the Planning Commission, made a perceptive observation as early as 1957. He wrote:

> It is a mistake to assume that shifting cultivation in itself is unscientific land use. Actually, it is a practical approach to certain inherent difficulties in preparing a proper seedbed on steep slopes where any disturbance of the surface by hoeing or ploughing will result in the washing away of the fertile topsoil. The tribal people therefore take care not to plough or disturb the soil before sowing. The destruction of weeds and improvement of tilth necessary for a proper seedbed are achieved with the help of fire (quoted in Bhowmik 1976: 8).

Chaturvedi and Uppal (1953) argue that the widely held notion that shifting cultivation is mainly responsible for large-scale soil erosion needs to be dispelled. Rather, it is to be seen as an agricultural practice that evolved in response to the physiographical character of the land (Bhowmik 1976: 7–8). They further state that a fairly well-distributed annual rainfall of more than 100 inches, high levels of humidity, and the peninsular character of rock formation provide the climatic and soil conditions conducive to the growth of vegetation that no sooner is a patch of ground left to itself than it is rapidly covered with a thick blanket of vegetation. The only time when the soil does get washed away is when it is under agricultural crops (Misra 1976: 82). What this indicates is that shifting cultivation, rather than being destructive, had a symbiotic relationship with the natural environment.

Cultural Dependence

The relationship between the natural environment and the tribes is not only woven around their physical existentce. It is evident in the cultural domain as well. Their myths and stories, riddles and puzzles, songs and dances, feasts and festivals, customs and life-cycle rituals all express, to a greater or lesser degree, various aspects of their relationship with the natural environment.

The natural environment assumes a very significant place in the creation stories of almost all tribes. The subject and the significance of these stories differ from tribe to tribe. This is seen most clearly in the Munda story of creation. The story runs as follows. At the beginning of creation, there was water everywhere. The Singbonga plied around aimlessly on a boat made of leaves. One day, he told his wife that he was fed up; there was no recreation. Therefore, he decided to create the earth and men. He commandeered the services of the fish, the tortoise, and the crab to bring up soil from the bottom of the waters, but they did not succeed. Ultimately, it was the leech that extended itself in such a way that its tail rose above the waters and its mouth sank into the soil underneath, its body serving as the channel for transporting the soil above the waters. So the earth was created, followed by the creation of plants, trees, animals, and birds. Among the birds was one called Hur, which produced a giant egg from which emerged the first two human beings, the ancestors of the Mundas.

The importance of the natural environment is also evident in the day-to-day religious beliefs and practices of tribal people. Spirits are believed to dwell in specific parts of the land, such as hills and hillocks, dense trees and vegetation, and water bodies such as rivers, tanks, and wells. The tribes hold or observe their feasts, customs, and dances to coincide with, and respond to, nature's shifting signals, such as the blossoming of plants and trees, the position of the moon, and on the changing seasons. For example, Sarhul, one of the most important festivals among the Oraons, the Mundas, and the Santhals, is celebrated when the sal trees are in full bloom, with branches of white flowers dominating the landscape. Nature and natural phenomena occupy a central place in rituals associated with this festival. Here a symbolic marriage between the supreme god (symbolized by the sun) and the mother goddess (the earth) is enacted. The ritual is performed with a view to ensuring the fertility of mother earth. A cluster of sal trees, which is considered a sacred grove, acts as a place of worship. The place and even the trees standing there assume importance.

The same is true of Karam, another important festival among these tribes. The tribes consider the karam tree to be sacred to the Karam deity, who has power to determine whether the autumn harvest will be good or bad. The festival is celebrated in September after paddy has been transplanted and has turned green. Boys and girls go to the forest to collect branches of the karam tree. They dance around the karam tree three or five times. Then a boy cuts young branches from a non-flowering karam tree. These are not allowed to fall to the ground while being cut and are handled with great care. Karam branches are then planted in the village courtyard. People sit around the karam branches and hear the story of the Karam rajas. After the storytelling session is over, the girls place flowers at the foot of the karam branches and dance around it the whole night. The next day, these branches are immersed in a pond or stream close by. The stories connected with the Karam festival indicate the tribal people's close bond with nature. These stories speak of punishment being meted out to those who treat the karam branches with disrespect. The punishment continues until the offender asks for pardon and the respect and honour due to the branches has been restored.

The major customs among tribes, as among other communities, are concerned with birth, marriage, and death. The link between customs and the natural environment is best reflected in customs connected with marriage and death. Among the Oraons, for example, as among other tribes, various pre-wedding customs are closely connected with the environment. Men go to the forest to fetch firewood and women go to fetch sal leaves for making cups and plates. The making of marriage mats and baskets of various sizes is another custom. The setting up of marwa is the most significant custom. Nine sal saplings with leaves on tops are planted in the courtyard in three rows. The middle of the second row differs in height. Branches of bamboo, sidha, bhelwa, mango, and mahua are also planted. The mango suggests perpetuity of descendants, the bamboo symbolizes progeny, the sidha stands for marital fidelity, the bhelwa offers protection from the evil eye, and the mahua symbolizes love between husband and wife. The marriage ritual would be incomplete without this invocation of trees and plants. Even the customs of disposing the dead vary depending on nature's signal. If a person dies after the new paddy seedlings have sprouted, which basically means the arrival of the wet or rainy season, then the Oraons follow the practice of burial. If a death takes place after the harvest and before the sprouting of the paddy, namely the dry season, then the body is cremated (Xaxa 1992: 105).

Customs Related to the Conservation of the Natural Environment

Since the dependence of the tribals on the natural environment is both subsistence-based and cultural, conservation practices among them indicate an obligation towards the environment, which is carried out in many ways. One of the important ways in which this obligation is enacted is through the system of totemism, a practice popular among many tribes in India, especially in eastern and central India. Totemism refers to the relationship between groups such as clans and various species of plants and animals. For example, among the Oraons, the Toppo clan takes its name from a bird, the Minj clan from a particular species of fish, and the Lakra from the tiger. Such is the case also among the Mundas, the Kharias, and the Hos. Although the principal use of totems lies in regulating marriage, it also serves other functions in that it orients people towards the environment in a particular way. Tribes observe certain taboos with respect to the objects of their totems. They do not eat, harm, kill, destroy, or even domesticate them. They do not use anything made or obtained from their totems. At times, such interdictions even extend to animals, birds, and plants that resemble either in shape or colour their particular totems. Thus, those who have the tiger as their clan totem may extend the attitude of reverence that they have towards the tiger to the squirrel as well, as the animals resemble each other with respect to colour and stripes. They do not believe that they have descended from their totems, but they do believe that these totems have helped, protected, or been of service to their ancestors in the past. Hence, they continue to maintain an attitude of reverence towards them. Totemism has resulted in many natural species becoming objects of respect for the tribes. The species that are taken as totems are not from any one single family of animals, birds, reptiles, or plants. Rather, where totemic practice exists, natural objects from all of these families partake in the construction of the social structure. The whole natural order thus becomes an integral part of the social order. Totemism not only expresses but also maintains the relationship of dependency between the social order and the natural order.

For tribals, the natural environment is also a dwelling place for many of their deities and spirits. These are believed to dwell in specific parts of the physical space, such as hills and hillocks, dense trees and vegetation, and water bodies such as rivers, tanks, and springs. Tribals in general respect and maintain these places. Of all the sacred places, the most respected and conserved is what is known as the sacred grove. A sacred grove is often a dense cluster of sal trees that is considered to be the abode of the guardian spirit of the village. Such attitudes and world views of tribes help to maintain and conserve the natural environment.

In addition, there are other taboos that regulate the relationship of tribes with their natural environment. Among the Oraons, for example, no man in the village is permitted to gather the new fruit, flowers, and edible roots of the season or to plough the fields before the feast of Sarhul, which, as mentioned earlier, marks the symbolic marriage of god with mother earth. Again, hunting expeditions follow certain taboos. Hunters do not kill or hunt any wild animals or birds during the months of June and July. They believe that the breaking of taboos can be harmful to the growing paddy crop.

The bonds between the earth and the tribes are not only material but also moral and ritualistic. Land is valuable to the tribes not only because it provides them the means of survival and livelihood but also because it was bequeathed to them by their ancestors. Hence, it is morally binding on the tribes to preserve the land and hand it over to their descendants. Through sacrifices and prayers, they seek to maintain a good relationship with their ancestors and the earth, on which their survival and livelihood depends.

Even subsistence-wise, their relationship with the natural environment is such that it helps to conserve nature. The community's dependence on nature, as noted earlier, is far from being passive. The community acts on nature and transforms it into forms that are of use to it. However, what is important is that this use of environmental resources is limited only to the extent that it is necessary for the community's survival. It is this attitude that leads to harmony between the community and the environment. Such harmony is possible because of the overriding social values that guide tribal societies. These are the values of balance between nature and culture, egalitarianism in social structure, collectivism in economic structure, accommodation in social relations, consensus in decision making, ethical living in philosophy, and group participation in music, dance, and art (Munda 1992). On these counts, the attitudes of tribes towards the world, including nature, is, to borrow the Weberian notion of rationality, one of rational adaptation and not one of rational domination, namely mastery over the world (Bendix 1960). The latter has invariably led to the exploitation of the natural world with a view to providing greater and greater comfort to humanity. However, provision of comfort at the expense of the natural world has posed immense danger to the very survival of the planet.

Political Economy of the Forest

The harmony between the natural and the social worlds that the tribes had maintained for centuries was ruptured during the colonial period. This had much to do with the forest policy of the colonial state. The policy had its genesis in a memorandum that outlined the rules for the conservation of forests in 1855. In a way, it was a blueprint for the forest policy of the colonial state that was adopted later in 1894. Prior to the introduction of the policy, tribals were virtually the lords of the forest, enjoying unrestrained access to forest and forest produce. The colonial policy, however, turned the situation upside down. It vested unlimited power in the state and paved the way for state control and management of forests. This led to a drastic curtailment of the rights and privileges that the tribals had hitherto enjoyed over the forest. In keeping with the newly conceived policy, the Indian Forest Act was drafted in 1865. It was revised in 1878 and made operational in most of the provinces. Laws relating to forests and forest produce were later consolidated in the Indian Forest Act, 1927 (Kulkarni 1983). The colonial forest policy was oriented more towards revenue and profit making than towards conservation. Hence, the reclamation of forests for land cultivation, expansion of roads and railways, exploitation of mineral resources, and industrial growth assumed the form of a key economic activity during the colonial period. Needless to say, these policies led to the large-scale destruction of natural resources. And yet the problems and hardships suffered by the tribals were not felt acutely until the post-independence period. Either the exact impact of the policy was not realized or the policy, despite expropriating the rights of tribal people, gave some leeway to them. That the latter was indeed the case is evident from the restrictions added to the New Forest Policy of 1952. The new policy, for example, introduced measures pertaining to the withdrawal of concessions on the release of forests for cultivation, withdrawal of the facility of free grazing, and weaning away tribals from the practice of shifting cultivation. Thus, under the new policy, certain concessions that the tribals had enjoyed during the colonial period were withdrawn. Further, there was now emphasis on increasing the area under forests as well as on measures aimed at stringent regulation, policing, and revenue earning. The policy also aimed to maintain one-third of the country's land area under forest. The drive to achieve this objective led to the claiming of even treeless land as forest land to be brought under the control of the forest department. Through processes such as this thousands of square kilometres of tribal land were brought under the forest department.

Interestingly, the new policy emphasized two goals—the need to increase land under forests, on the one hand, and the need to utilize forests for national needs and maximization of revenue, on the other. National needs entailed the development of infrastructure, acceleration of industrial and agricultural growth, and increased exploitation of minerals. The private accumulation of profit by the Forest Department in collusion with contractors has been the other important factor. This is not to say that tribals have had no hand in it. The loss of livelihood, either because of displacement or for other reasons, has forced many of them to cut trees and sell them in the market as a part of their survival strategy. All these practices have depleted existing forest resources. Paradoxically, most development projects in post-independence India have taken place in tribal-inhabited areas, which comprised large tracts of forest land. In short, it hardly needs any reiteration that the destruction of forest resources has been caused more by development projects and state revenue demands than by tribal needs. Not only that, even the benefits of such development projects have not accrued to the tribals. On the contrary, they have been thrown off the land and forests that had been their life-support system for centuries. There has not even been adequate compensation, let alone rehabilitation, for the land that they were forced to part with for the sake of national development. That tribals have paid a heavy price on account of this development is evident from the fact that while they comprised only 7.5 per cent of the total population of the country, they constituted 40 per cent of the displaced population by 1991 (Planning Commission 2001). Yet a section of the cosmopolitan environmentalists and conservationists tend to accuse tribals of putting the country's forest wealth at stake.

What is even more intriguing is that forest resources, despite being under state control and management, have undergone serious depletion throughout the decades. In the thirty years since the declaration of the policy, in 1952, 4.3 million hectares of forest land were diverted to non-forest purposes (Kashyap 1990). Thus, state control and management of forest resources has not improved forest conservation in any way. Rather, the policy has had the reverse effect. Hence, in order to provide impetus to the protection and conservation of forests, the Forest (Conservation) Act, 1980 was adopted. The Act forbids measures such as de-reservation of forest lands, use of forest lands for non-forest purposes, clearing of forest trees, etc. by state governments or other authorities without the prior approval of the Union Government. The Act has proved to be instrumental in controlling the indiscriminate diversion of forest land for non-forest purposes. Since the adoption of the Act, the rate of such diversion has fallen to a modest 16,500 hectares per year from the earlier figure of about 1.5 lakh hectares per year (Kashyap 1990: 316). The Act has had far-reaching implications for tribals. In freezing legal land use of lands officially recorded as forest through the process referred to above, the Act declared millions of tribal forest dwellers illegal occupants on their own lands. Not surprisingly, it has led to a great deal of disquiet and unrest among tribal people. In response, the Ministry of Environment and Forests (MOEF) issued several circulars aimed either at the regularization of encroachers or at the resolution of disputed claims, but the progress made on this front was insignificant. Meanwhile, India adopted a New Forest Policy in 1988, which for the first time acknowledges the symbiotic relationship of the tribals with forests and the rights and privileges that they have traditionally enjoyed over forests. The policy hence envisages protecting the interest of tribals (Verma 1990: 191). The national policy of 1988 even makes provision for regularizing what it calls tribal encroachments, and it takes 1980 as the cut-off year for this purpose. Despite such well-meaning provisions, the policy has failed to extend any tangible benefits to tribal and forest people. Much of this problem is the result of the judicial pronouncements on the Godavarman PIL before the Supreme Court.

The judicial pronouncements on the case have extended the ambit of the 1980 Act even to lands that have yet to be finally notified under the Indian Forest Act and to all lands conforming to the dictionary definition of forests, irrespective of ownership. The pronouncements had even stayed the regularization of eligible pre-1980 so-called encroachers (Sarin 2005). As part of efforts to enforce the court order, the MOEF has embarked on a drive of mass eviction of tribal and other forest dwellers from lands they have been occupying but that have been declared as forests by the forest department. This is the context in which the scheduled tribes (Recognition of Forest Rights) Bill, 2005 should be seen. It incorporates the provisions made in the National Forest Policy. It recognizes the right of forest-dwelling scheduled tribes to forest land under their occupation for habitation or self-cultivation for livelihood needs. The right to forest land is in no case to exceed 2.5 hectares and is subject to the condition that such forest-dwelling scheduled tribes have occupied the land or have acquired forest rights before 25 October 1980.

The scheduled tribes (Recognition of Forest Rights) Bill, 2005 seeks to correct the 'historic wrong' whereby the traditional forest rights of the tribals had been denied under British rule. The draft bill notes that even after independence, India continued to follow the colonial forest policy, dispossessing the tribals even further. These errors were now sought to be rectified by giving a permanent stake in forests to the scheduled tribes living in forests and by associating them with forest protection and conservation. To Guha (2005), this philosophy of conservation represents a significant shift from the policy followed in India for more than a century, which held that forests could be protected only by designating local communities as interlopers and enemies. The Bill, however raised a storm of protest from a section of powerful cosmopolitan environmentalists and the wildlife conservation lobby, whose members believe that giving tribals their long overdue rights will deliver a body blow to attempts aimed at environmental conservation and wildlife protection. There is no denying the truth that there has been widespread destruction of wildlife in India. However, this is not the result of the activities of tribals but the outcome of indiscriminate hunting by rajas and zamindars, ruthless destruction of the forest habitat by the state to meet its revenue demand, and commercial and industrial exploitation. Environmentalists and conservationists, however, see the danger to environmental conservation and wildlife protection emanating primarily from the tribes without even being able to substantiate the extent of wildlife destruction caused by tribal people. Anyone who is familiar with tribal life and culture knows that this is not the tribal ethos.

Forests have been mercilessly and thoughtlessly denuded during the last fifty years. Between 1951 and 1972, 3.4 million hectares of forested area were cut off for use in agriculture, industry, dams, roads, and so on (Shiva et al. 1983). The present rate of deforestation in India is still an incredible one million hectares every year. These statistics do not include the areas near dams and industries where tree cover has been completely destroyed. They also do not include illegal felling and contract felling by the Forest Department. However, interestingly, for the environment and wildlife lobby, the damage caused to forests by these projects has gone unnoticed. These lobbyists did not raise a hue and cry over the loss of forest wealth. Hence, one could hardly have expected them to notice the fact of mass displacement of tribal people resulting from these projects. Yet the loss of forest cover (even when it has no forest vegetation) resulting from the extension of rights to tribals for their livelihood security is immediately noticeable, and a hue and cry is raised accompanied by a distortion of facts and figures. It is worth noting that the rights extended to tribals are not something new. The rights are only being restored. These rights, which they had traditionally enjoyed for centuries, were unjustly taken away from them during the period of colonial rule. Post-independence India continues to perpetuate this injustice, and even expand this process.

Forest Department statistics show that 23 per cent of India's total area is forest land. But this is not forested area. It simply means that this area is under the control of the Forest Department. The National Forest Policy of 1952 recommended that 33 per cent of the country's area should be brought under forest cover. Between 1951 and 1988, under the provisions of the colonial periods Indian Forest Act, 1927, the national forest area was increased by another 26 million hectares, that is, from 41 to 67 million hectares (Sarin 2005). In an attempt to achieve this target, the Forest Department, as was done in the case of the family planning project, went on a spree to bring non-private lands under its control through blanket notification, without even surveying the vegetation or ecological status of these lands and without recognizing the rights of pre-existing occupants and users as required by law. Tribal areas bore the brunt of this drive to achieve statistical success because of poor record-keeping and documentation of the adivasis' customary rights.

Does the onus of maintaining 33 per cent of India's land under forests lie only with the tribals? As it is, 60 per cent of state forests today are concentrated in 187 tribal districts confined to only one-third of the country. In the process, many of the most vulnerable tribes have been stripped of their customary resource rights without even their knowledge and have been labelled encroachers on their ancestral lands (Sarin 2005). The tribals thus have contributed much more than their fair share to the conservation of forest wealth in India. If tribals have been made to sacrifice their rights over land, forests, and other resources for the sake of maintaining the ecological balance of the country for the last hundred years, the onus now lies on the rest of India to step forward, because the contribution of non-tribals to maintaining the ecological balance of the country has been negligible. The non-tribals have been the beneficiaries of this system over the decades. Even in the tribal-inhabited hill regions of the north-east, where shifting cultivation is an age-old practice, and whose inhabitants are seen as threatening forest wealth, forest wealth is far more extensive than in the rest of India. Of course, the land available for shifting cultivation has now shrunk because of various kinds of state- and non-state-sponsored development activities. This factor has affected shifting cultivation insofar as the duration of the cultivation cycle has now been shortened, which in turn has affected soil fertility and productivity. Nevertheless, the tribals' contribution to sustaining forest wealth far exceeds that of the inhabitants of the rest of India. In fact, where state control and management of forests has been historically absent, there has been less destruction of forest wealth. The hill regions of the north-east, where state management of forests was introduced under the forest policy of 1952, bear testimony to this. The total forested area in the region in the 1980s, including the plains, constituted about 53 per cent of the total reported area. Of this, 36 per cent was reserved and 64 per cent was unclassified and thus open for use by local people. If one were to exclude the plains (the forests in the plains had been under state control and management since the colonial period, yet there has been greater damage of forest wealth there), the area under forest cover in the region would increase further (Xaxa 1998; Patnaik 1987). Thus, the contribution of the north-eastern hill regions, popularly known as the tribal states, to India's forest wealth far exceeds the contributions of the other states.

Since everyone gains from the maintenance of the ecological balance, all citizens—not tribals alone—should be responsible for achieving this goal. Equitable distribution of responsibility demands that non-tribal states and districts should find out how much of their land is under forest cover and how much they have to contribute to maintain national standards. If tribals could be dispossessed and displaced, and even suffer livelihood problems, to maintain the ecological balance of the country, there is no reason why non-tribals and non-tribal states should be spared from going through a similar process.

References

Bendix, Reinhard. 1960. *Max Weber: An Intellectual Portrait.* London: Heinemann.

Bhowmik, P. K. 1976. 'Shifting Cultivation: A Plea for New Strategies', in *Shifting Cultivation in North-East India: Seminar on the Socio-Economic Problems of the Shifting Cultivation in North-East India with Special Reference to Meghalaya*, edited by B. Datta-Ray. Shillong: North East India Council for Social Science Research, pp. 6–13.

Chaturvedi, M. D. and R. N. Uppal. 1953. A Study in Shifting Cultivation in Assam. New Delhi: Indian Council of Agricultural Research.

Fernandes, Walter. 1993. 'Forests and Tribals: Informal Economy, Dependence and Management Traditions', in *Continuity and Change in Tribal Society*, edited by Mrinal Miri. Shimla: Indian Institute of Advanced Study, pp. 48–69.

Guha, Ramachandra. 2005. 'Tribal Pursuits.' *Hindustan Times*, New Delhi, 16 June.

Hebbar, Ritambhara. 1995. 'Forest-Tribe Affinity: The Case Study of Bastar (Madhya Pradesh)', M.Phil Thesis, Department of Sociology, Delhi University.

Hoffman, Johan B. 1950. *The World of the Mundas.* New Delhi: Critical Quest (first published in the *Encyclopaedia Mundarica* (1950)).

Kashyap, Subhash C. (ed.). 1990. *National Policy Studies.* New Delhi: Published for the Lok Sabha Secretariat [by] Tata McGraw-Hill Publishing Company Ltd.

Kujur, Alomani Anupa. 1989. *The Oraon Habitat: A Study in Cultural Geography.* Ranchi: Daughters of St. Anne.

Kulkarni, Sharad. 1983. 'The Forest Policy and Forest Bill: A Critique and Suggestions for Change', in *Towards a New Forest Policy: People's Rights and Environmental Needs*, edited by Walter Fernandes and Sharad Kulkarni. New Delhi: India Social Institute, pp. 84–101.

Misra, B. P. 1976. 'A Positive Approach to the Problem of Shifting Cultivation in Eastern India and a Few Suggestions to the Policy Makers', in *Shifting Cultivation in North East India: Seminar on the Socio-Economic Problems of the Shifting Cultivation in North-East India with Special Reference to Meghalaya*, edited by B. Datta-Ray. Shillong: North East India Council for Social Science Research, pp. 106–121.

Munda, Ram Dayal. 1992. 'In Search of Tribal Homeland', in *Tribal Transformation in India.* Tribal Studies of India Series, Vol. 3, edited by Buddhadeb Chaudhuri. New Delhi: Inter-India Publications, pp. 377–86.

Patnaik, S. C. 1987. 'Dependent Economy of North East Region of India and Alternative Planning Techniques.' Paper presented at the National Seminar on Institutional Economics, Shillong.

Planning Commission. 2001. Report of the Steering Committee on Empowerment of the Scheduled Tribes for the Tenth Five Year Plan. New Delhi: Planning Commission, Government of India.

Sarin, Madhu. 2005. 'Commentary–Scheduled Tribes Bill 2005: A Comment'. *Economic and Political Weekly*, 40 (21): 2131–34.

Shah, Mihir. 2005. 'Governance Reform for India's Forests', The *Hindu*, New Delhi, 20 May.

Shiva, V. et al. 1983. 'The Challenge of Social Forestry', in *Towards a New Forest Policy: People's Rights and Environmental Needs*, edited by Walter Fernandes and Sharad Kulkarni. New Delhi: India Social Institute, pp. 48–72.

Verma, R. C. 1990. *Indian Tribes Through the Ages*. New Delhi: Publications Division, Ministry of Information and Broadcasting, Government of India.

Xaxa, Virginius. 1992. 'Oraons: Religion, Customs and Environment', in *Indigenous Vision: Peoples of India, Attitudes to the Environment*, edited by Geeti Sen. New Delhi: Sage Publications, pp. 101–110.

———. 1998. 'Cultural Dimension of Ecology: A Case Study of the Oraons', in *The Cultural Dimension of Ecology*, edited by Baidyanath Saraswati. New Delhi: Indira Gandhi National Centre for the Arts, pp. 125–28.

Chapter 10

Women and Gender: Aspects of Inequality in Tribal Society

Ethnographic accounts of tribes in the form of monographs, handbooks, and inventories have been the hallmark of tribal studies in India until the country attained independence. These provide detailed information on different aspects of life among the tribes, such as economic activities, marriage and divorce, family relations, kinship patterns, religious practices, technology, arts and crafts, customs and traditions, rules of inheritance and succession, division of labour, and so on. It is from these accounts that we get an idea of the position of women in tribal societies, the restrictions and taboos they suffer, and the rights, privileges, and freedoms they enjoy.

Tribal Women in a Traditional Setting

Studies on tribal women have been few, and these lack depth and comprehensiveness. In 1987, the Indian Anthropological Society published a region-wise survey of tribal women. The survey points out that the subject of tribal women has generally been ignored, and when studies do exist, they are brief or disposed of summarily.[1] Except in some of the works of Elwin, Furer-Haimendorf, and some others, there has been no explicit discussion of the status of women in tribal society. Even in these works, the assessment of the status of women has been far from uniform. While some say that tribal societies generally assigned high status to women, others argue that women's position in tribal society is the same as that in other societies. Writing about the Nagas, Elwin, for example, remarks that 'tribal woman is in herself exactly the same as any other woman, with the same position, love and fears, the same devotion to the home, to husband and children, the same faults and the same virtues' (cited in Zehol 1998: 1). Yet elsewhere, he refers to Naga women as holding a high and honourable position, and describes them as working on equal terms with the men in the fields and making their influence felt in the tribal council (Elwin 1961: 104). He makes a similar observation with regard to the Baigas, noting that Baiga women enjoy an excellent position in society. They enjoy freedom and exercise authority, play a leading part in marriage ceremonies, and have a reputation for witchcraft. Among the Baigas, there is no clear division of labour between men and women (Elwin 1986: 235–236). Furer-Haimendorf referring to the Nagas, on the other hand, writes:

Many women in most civilized parts of India may well envy the women of the Naga Hills their high status and their free happy life, and if you measure the cultural level of the people by the social position and personal freedom of its women you will think twice before looking down on the Nagas as savages (Furer-Haimendorf 1933: 101).

In a similar vein, Hutton attributes a higher social status to Sema Naga women on the ground that marriage among the tribe is based on choice and a girl is never married against her will. The Sema Naga woman occupies a high position in her husband's house and is treated well (Hutton 1921: 183). Other accounts, barring a few exceptions such as those by Mann (1996), Chauhan (1990), and Zehol (1998), are limited to a few paragraphs, or at best to a page or two, but with a discussion on the status of women largely missing.

The discrepancy in the descriptions of women in tribal society is mainly due to the way in which scholars conceive the term status. Broadly, two views may be discerned. The first refers to women's role in the social system, which entails rights and duties. Here status is linked with the role of women in the social system. Women's status is analyzed and assessed taking into consideration the mutual duties shared by men and women and the safeguards provided for the protection of each sex against the highhandedness of the other. The other usage of the term status is in the sense of prestige and honour, which may be studied in terms of women's legal status and opportunities for their participation in the wider society. In the studies of tribal women, however, such a distinction is often overlooked, and is almost used interchangeably.

Status in terms of prestige and honour can only be studied in relation to values in reference. Tribes have been studied primarily either with reference to the values of the Enlightenment, viz. freedom, equality, and fraternity, or in relation to the values prevalent in the dominant society. In short, in studying tribes in relation to the 'other', the values have invariably been either those of the universal values of the Enlightenment or the values of the larger society in which freedom is often scorned and hierarchy much valued. Hardly any attempt has been made to study tribes in terms of the values prevailing in tribal society.

Women in a Changing Tribal Society

Sociologists and social anthropologists have questioned the dichotomous conception of Indian society constructed during the colonial period. Though a distinction is maintained between 'tribe' and 'civilization', the two are not treated as isolated but as being in interaction with each other (Sinha 1958). Consequently, tribal society has not been seen as static but as undergoing a process of change. Change has been conceived as moving in a number of directions. Of these perspectives, the dominant view has focused on the change from tribe to caste. Indeed, much of the social anthropological discourse on tribes has been couched primarily in terms of the transformation of tribes into castes.

Changes occurring in tribal society invariably have been conceived of in terms of the tribe moving in the direction of becoming a part of 'civilization', by being assimilated into the society that the civilization represents. Historians and anthropologists have made such observations not only in the context of the past but also in the context of the present. Kosambi refers to tribal elements being fused into the general society. Similarly, Bose refers to tribes being absorbed into Hindu society. The fact that such claims have continued to appear in a large number of anthropological works in the post-independence period points to the persistence of the belief that tribes are being absorbed or assimilated into Hindu society, or that tribes are gradually turning into castes. In this journey towards absorption or assimilation, tribes are said to be at different levels or stages. Nowhere is this better reflected than in the classifications adopted by social anthropologists attempting to understand the process of tribal transformation in India. Tribes have not only moved in the direction of Hinduism and the Hindu way of life. Equally important has been their movement towards Christianity. All these developments have been considered significant processes of cultural change in tribal societies. Yet we know little about the way these changes have affected the lives of women in these societies. How have women whose families have been acculturated into Hinduism adapted to the ethos of the new religious organization? What has been the extent of continuity and discontinuity in their lives? What have been the advantages and disadvantages faced by women in these changed situations? These questions have not been explored systematically in existing studies.

Some studies on the Sanskritization process among tribes do point to some changes. Roy Burman, drawing on the studies made during the period from the 1930s to the 1960s, demonstrates how tribes undergoing the process of Sanskritization are opting for early marriage as a matter of prestige and are discouraging widow remarriage as well as divorce and separation (Roy Burman 1988: 14; Sachchidananda 1988: 80). Mann makes similar observations regarding the Bhils. The freedom enjoyed by Bhil women in the spheres of marriage, premarital sex, divorce, access to decision-making bodies, etc. is being gradually curtailed with the adoption of Sanskritization. Their freedom in selecting a mate and marrying through elopement has been adversely affected. The purdah system among Bhil women is again an instance of borrowing from caste Hindus (Mann 1987: 155). Roy Burman regards divorce and separation in rural or tribal India evidence of desertion, indicative of the powerlessness of women. In this sense, he views Sanskritization as a positive influence. Further, he tends to link age at marriage more with the customary practice of bride price than with Sanskritization or distance from it. Citing the case of the Hos of Chhotanagpur, he points out that it is not uncommon among them for a woman to wait until her mid-thirties before she could find someone who could pay the bride price and become her life partner (Roy Burman 1988: 14).

With improvement in economic conditions, or even Sanskritization, it is argued, women workers tend to withdraw from outdoor agricultural work and allied activities. Observations available in the existing literature on the inverse relation between the rise in social status and the withdrawal of women from manual labour are taken in support of such an argument (Roy Burman 1988: 16). These observations, however, are brief, casual, and lack analytical rigoar. To begin with, Sanskritization is not the only basis for the rise in social status in tribal society. There is no doubt that with Sanskritization there has been a tendency among tribes to withdraw women from manual labour, but this has more to do with well-to-do cultivators who either inhabit multi-caste villages or their vicinities. Withdrawal of female labour from contexts and situations other than the one referred to above, even when Sanskritization has been at work, is usually absent. In other cases, rise in status, such as that resulting from a change in occupation, especially white-collar employment, has not necessarily led to the withdrawal of women from manual labour. Indeed, where such individuals are also tied to land and agriculture, women have continued performing their routine gender-specific agricultural activities. Where male members of the family are compelled to live away from the village because of the demands of employment, women have borne the added responsibility of organizing agricultural activities usually performed by men.

Whereas acculturation processes such as Sanskritization and Hinduization led to change of a certain kind among tribal women, the change moved in a somewhat different direction in the context of the conversion to Christianity. Christianity opened up the space for tribal women to participate in religious worship side by side with men, something that had been denied them under the earlier religious traditions of many tribal societies. Many Protestant denominations allow women to be appointed as deaconesses, preachers, and trustees. They have also been organized into women's groups within churches for carrying out social, religious, and welfare activities (Kelhou 1998: 59–60; Zehol 1998: 26–27).[2] Christianity as practiced by these groups has also made modern education accessible to women in tribal societies. On the other hand, it also introduced a variety of restrictions in the name of church ethics and law that militated against the freedom women had enjoyed in traditional tribal societies. Christianity also introduced segregation among boys and girls and put a stop to divorce, which was easy to work out in accordance with tribal customary law. Over and above this, gender inequality inherent in Christian churches and denominations continued.

Both Hinduization and Christianization thus led to a number of restrictions on the kinds of freedom women enjoyed in the traditional social setting. In the case of Hinduization, such restrictions were part of the concern with maintaining respectability and status; in the case of Christianization, it had more to do with religious morals and values. An inherent inadequacy of this perspective is that it fails to capture the variety of changes that occurred in tribal societies.

Assumptions Underlying the Study of Tribes

Underlying the accounts of tribes in general and of women in particular is the basic assumption that tribes are primitive, savage, and backward. This point is made repeatedly in accounts of their means of livelihood and survival, technology, food habits, lifestyle, and, more importantly, through representations of their bodies. Tribes were invariably represented as half-clad, clothed in only leaves and grass, and at times as naked. They are shown vividly in pictorial form in anthropological and other works to emphasize their primitive existence. Such pictorial representation is now no longer the special privilege of anthropologists. It has now become an object of commerce, as seen from the sale of picture postcards of Andamanese tribes like the Jarwas and the Onges. Women feature more prominently than men in these representations. Some ethnographers like Elwin and Furer-Haimendorf celebrated such practices among tribes. Elwin described in detail the place of sex in tribal life, discussing the sexual life of children, the notion of erotic and sexual attractiveness, courtship and romance, and frequency of intercourse, for example, in the case of the Baigas (Elwin 1986: 230–270). That these practices were governed by certain codes and ethics has been overlooked or under-stressed.

While some anthropologists celebrated the status of women in tribal society take the states of their women. others considered tribes as savages and ridiculed them precisely on grounds. They criticised their character, customs, and morals. Tribal women had an exaggerated place in such representations. The dominant view of tribes and tribal societies treated them socially and culturally as almost animal-like. A study, for example, shows that as many as 95 per cent of upper-caste women respondents consider Bhil women socially inferior (Mann 1987: 105). Some administrators and scholars who wrote on tribes and tribal life also perpetuated such a view. This view was even internalized to some degree by members of the tribes themselves. Tribes were so viewed and treated precisely because their values were contrary to the dominant values of the larger society. Mandelbaum puts it this way:

> Tribal people generally take direct, unalloyed satisfaction in pleasure of the senses, whether in food, alcoholic drink, sex or dance. The twice-born tend to be ambivalent about such pleasures, they are inclined to defer them or refine them and surround them with elaborate ritual (Mandelbaum 1970: 583).

The values governing the larger Indian society are, however, in the process of changing. Indeed, there has been a paradigm shift with regard to the nature of values sought and at work in the post-independence period. The ethos in the period as embodied in the Constitution and the country's laws is one of freedom, equality, and social justice. The latter principle also features prominently in contemporary public rhetoric. Along with this there has been a shift in the image of women in tribal society, especially among the educated and the enlightened. Ethnographers, British as well as Indian, were on the whole rooted in the spirit of Enlightenment rationality. To them, freedom, equality, and fraternity were the basic values in the assessment of individuals and groups. They invariably judged the position of women in tribal society in the context of the dominant values of the West, which were contrary to the dominant values of the larger Indian society. Therefore, almost all discussions on women in tribal society were pitched against the position of women in a caste-based society.

In recent years, there have been more focused discussions on the status of women in tribal society, and yet these discussions have also moved in different directions. Women in tribal society were portrayed as having a better status than that held by women in caste-based societies. Tribal women were shown as having greater freedom with respect to movement outside the home, choice in marriage, divorce, and remarriage, and access to property and resources. This is revealed by anthropological studies on the rules of inheritance, right to property, enjoyment of freedom, share in the decision-making process, etc. In short, the studies focus on the rights and privileges enjoyed by women, on the one hand, and the roles and responsibilities assigned to them because of their sex, on the other. Hence, tribal women were invariably depicted as having a higher social status than their counterparts in caste-based society. These aspects of tribal society are still highlighted and are further corroborated by such demographic evidence as sex ratio and female workforce participation. This is the general image of women depicted in studies on tribal society. Nowhere is this more evident than in the proceedings of a series of seminars organized by the North-East Regional Centre of the ICSSR in different parts of the region on the status of women, held between May 1988 and February 1989.[3] Topics such as the economic burden and workload borne by tribal women as well as their access to education, food and nutrition, modern occupations, and political participation, especially in the modern context, have not received the kind of attention they deserve.

Stages of Social Formation

The position of tribal women in the context of all-round socio-economic changes in tribal society has been another area of focus in the study of women in tribal society. Their position, involvement, participation, and role in various aspects of tribal life are discussed and analyzed in the context of such wider socio-economic changes.

One of the dominant ways of looking at change from this angle is to study changes in the means of livelihood and survival. This is seen most clearly in the change from food gathering to food production, or from swidden (slash-and-burn cultivation) to settled agriculture. Such a transformation is also seen as a shift from communal and collective ownership of land and use of labour to private ownership of land and labour.

Developments such as these have led to a critical examination of the idea of tribal society as egalitarian. Forms of inequality in traditional tribal society have been brought under scrutiny. Of these, gender inequality, has been highlighted as the most pervasive, irrespective of the stage of social formation. One of the aspects that has been highlighted in this context has been the relative position of women and men under different types of social formations such as food gathering and hunting, swidden agriculture, settled agriculture, and state formation

(Nathan 1997; Menon 1993). The other way in which the position of men and women has been studied is in the context of different tribal societies without useing their social formations as the reference point. This has been done with respect to division of labour, forms of property, religious institutions, family organization, and the state. Through an examination of the former, an attempt has been made to show how the establishment of patriarchy took place in tribal societies. It is shown how in social formations such as food gathering and swidden agriculture, a rigid division of labour was either absent, or how gender inequality in one sphere was offset by equality in another sphere. Citing the case of the Birhors of Jharkhand, Nathan, for example, shows how the greater importance of men due to their participation in the public domain, in this case, distribution of meat (prestige food) throughout the society, is neutralized by a similar kind of engagement by women through exchange and transaction activities in the market. Similarly, among the Khasis, the higher social status of women due to rights of ownership held by them over ancestral property is neutralized by the men's control over the decision-making process. It is not clear whether this inequality that Nathan explicates has more to do with shifting cultivation or with the institution of matriliny and settled habitation, as in the case of the Khasis. Again, Nathan primarily traces the monopolization of upper levels of the hierarchy by men in the formation of the state and to the establishment of individual property. Men's control over the ritualistic and public or political spheres is seen as a crucial factor in the struggle to exclude women from ownership of land (Nathan 1997). Yet the cases of the Khasis with state-like institutions, and the Jaintias with full-fledged state institutions, do not seem to support such an argument. Women continue to own land, and the monopolization of upper levels of the hierarchy is still absent among them.

Indeed, some of the serious problems from which women suffer, such as being accused of witch-hunting, are traced by scholars to the pattern of landownership in tribal societies. Kelkar and Nathan, for example, argue that it is the life interest of a widow in the entire land of her husband that is a major fetter on the property rights of the husband's male agnates. Following this, they argue that victims of witch-hunting are not only women in general, but widows who have a life interest in their husband's land in particular. The life interest of a widow in her husband's land restricts the property rights of the male agnates of the deceased, for whom the use of the land for accumulation or for consumption has to wait until after the death of the widow. A widow without children is, therefore, more vulnerable to attack on the pretext of being a witch (Kelkar and Nathan 1997: 263). While there may be some association between the two in some cases, it is difficult to generalise that witch-hunting is connected primarily with ownership of land. Anyone concerned with the study of witchcraft is familiar with the fact that it is much more complex than a practice associated only with property rights. Indeed, a number of witch-hunting cases have been reported from time to time from the tea gardens of Jalpaiguri district in West Bengal, where access to property was hardly the issue (Gupta 1979).

The myth of gender equality or the higher status of women in tribal societies has also been brought under scrutiny through the examination of such topics as customary law in relation to women in the areas of property, inheritance, marriage, and so on. It has been shown that women in tribal societies are put at a disadvantage vis-à-vis men (Nongbri 1998). It is interesting to note that the very practices illustrated as pointers to higher social status in one kind of setting turn out to be in-built depressors in another. Nongbri makes this point in her discussion of bride price. Referring to several tribes in Arunachal Pradesh, she points out how women among them are treated as mere commodities, procured easily by men through financial settlement in the form of bride price, a custom originally intended to compensate the girl's family for the loss of an economically active member. The custom of bride price as the socially legitimate purchase of a wife has thus become a convenient justification for men to abuse their wives and to treat them as disposable commodities. Nongbri writes that the system of bride price has proved to be a bane for women and lies at the root of the proliferation of polygamous unions as wealthy men can take on a number of wives simply by paying an agreed upon sum to the girl's family (Nongbri 1998: 33–34). Unnithan-Kumar in his study of the Taivar Girasia, a tribal group in Rajasthan, points out that bride price is viewed not so much as a recognition of a woman's contribution to the household or as a payment for the loss of a productive member, but as compensation to the father of the bride and his agnatic group for past expenditure on her maintenance, particularly consumption of food. Based on this, Unnithan-Kumar argues that bride price (valued in practical terms) is regarded as an important contribution that women make to the production of reproduction. The lack of ownership of property rights by women is justified by the Girasias on the ground that women after marriage move from their natal to their affinal villages (Unnithan-Kumar 1997: 205–206).

In this context, there are two aspects on which I would like to comment.

(1) It is an established fact that the division of labour in tribal society is based more on sex than any other factor. Division of labour has been both fluid and rigid. It is, however, not clear as to what stage and on what principle the division of labour in tribal society could be said to have assumed the form of inequality of rank or status between the sexes. There is little discussion of how and on what basis the differences, especially those between the sexes, came to be graded (whether divisions were merely the those of work and labour, and therefore devoid of the evaluation and gradation, intrinsic to the evaluation of being of high or low social status). Hence the study of the status of tribal women poses certain problems. It is difficult to study women from the perspective of the values inherent in these societies, especially since these values are based on notions of collectivity, equality, and sharing. Hence, rather than talking of high or low social status, it is more pertinent to talk of gender inequality. In the case of the latter, one can examine the relative position of women and men in relation to their access to equal opportunity, both formal and substantive.

(2) The taboo on women touching or using the plough in tribal society has been seen as a way of denying women control over the means of production, viz. land (Nathan 1997; Kishwar 1987: 96). This lack of access to and control over land is, however, already denied to women in tribal societies by existing customary laws. Oraon and Ho women, forbidden by taboo from holding the plough, for example, are already denied access to land by the customary laws of their societies. The explanation for the denial of women's access to land in terms of taboo thus seems far from adequate. After all, even the Brahmans are forbidden by taboo from ploughing, but that does not mean denial of access to land for them. Even in swidden agriculture, which Boserup describes as a women's farming system, the allocation of plots is made to men in their capacity as the heads of households, though women exercise greater control over the plots after they have been allocated. There has also been tribal land under swidden agriculture controlled and distributed by the chiefs, such as among the Mizos, the Semas, and the Konyaks. There is no custom of plough cultivation among these tribes; yet women are denied access to land even in these communities.

Social Differentiation

Another way of locating change has been in the increasing social differentiation taking place in tribal societies. Such social differentiation is rooted in forces outside of tribal societies, such as introduction of private property in land, growth of trade and the emergence of the market, emigration of non-tribes in search of land and employment, spread of modern education, opening up of new occupations, state-sponsored programmes, and similar activities. Accordingly, social differentiation has moved along different lines. The tribes lost their land and were compelled to take up employment as labourers in nearby quarries and coal fields and in emerging towns as unskilled or semi-skilled workers (Vidyarthi 1970; Banerjee 1981), or move permanently or temporarily elsewhere in search of work, such as plantations in Bengal and Assam. A minority has been able to take advantage of the forces unleashed by the market and the benefits extended by the state to tribes. This has led to differentiation among tribes based on criteria such as education, occupation, income, wealth, and assets. They have become differentiated into categories such as the rich, the middle, the poor, and the landless. This has led to the emergence of class relations, traditionally absent in tribal societies (Shah 1982; Bose 1984; Pathy 1984). There is, however, a lack of a gender perspective on the differentiation witnessed by tribal societies over the years. There has been no sustained discussion on whether relations between men and women have remained the same across different social categories in tribal societies or whether they have changed. A few studies do provide some insights into this dimension. For example, Punalekar in his study of tribes like the Dhodia, the Chodhra, the Gamit, and others in Gujarat points to a division between the well-to-do and the lower strata. This difference in social and political behaviour is also reflected in the social situation of the women of the two groups. Women of the well-off section have come to enjoy certain advantages. There daughters are enrolled in local schools for pre-primary education. There is a distinct tendency among these sections to provide girls higher education in schools away from the native village. Many of them join professional courses, such as teaching and nursing, and some even pursue technical courses. They also take advantage of the facilities provided under the reservation policy. Migration to cities for education provides them with opportunities to acquaint themselves with urban ways of living and thinking. Their food habits and dress preferences reflect the influence of the urban way of life. There is also a tendency among them, on the one hand, to emulate the norms and practices of the women of the dominant caste groups and, on the other hand, to deliberately abandon or de-emphasize their own traditional customs, rituals, and social practices. There is also increasing disregard for, and even indifference towards, women of the lower stratum of their tribe or village.

In contrast, tribes from the lower stratum have low school enrollment rates and high dropout rates. Their children hardly move beyond the primary level. Hence, they are constrained to work as construction labourers, domestic servants, cart pullers, scrap collectors, vendors, etc. In this struggle for survival, women make a valuable contribution and face considerable hardships. Yet their role and participation under these difficult conditions is hardly recognized. Further, due to such struggles, daughters are forced to discontinue their studies and go out to earn a living to supplement the family income. In towns, women develop a heightened sense of insecurity and dependence. Indeed, often their fathers, husbands, or brothers decide matters concerning their work and wages. Often, such decisions taken by the men are challenged if they are taken without consulting the women or at lower wage rates (Punalekar 1988: 94–102). Sachchidananda shows that educated Hos in white-collar jobs in recent years aspire for homebound wives as a marker of higher social status. This ultimately has lowered the status of women (Sachchidananda 1988: 84).

The discussion of bride price also reveals glimpses into the kinds of changes taking place, especially among the richer sections of tribal society. The rich among some tribal communities of Arunachal Pradesh have been using the traditional practice of bride price for their own sexual aggrandizement (Nongbri 1998; Mann 1988). There has also been a shift from bride price to dowry among the educated and salaried sections of some tribes (Misra 1984: 107; Kishwar 1987: 151).

Tribal Movements

Tribal studies have moved in many directions, with the major focus currently on development issues and tribal movements. However, the study of women in tribal society has remained confined to a discussion either of status or of inequality. One of the most important features of tribal studies has been the study of tribal movements. Besides contributing to ethnography, the studies of tribal movements have given rise to certain kinds of generalizations or conceptualizations about such movements. One of the general concerns in the study of tribal movements in India has been to provide a system of classification. Generally, tribal movements have been classified on the basis of factors such as geographical distribution, chronology, aims and objectives, and issues around forest, land, ethnicity, political autonomy, secession, mobility, and identity.

The study of movements from the perspective of social strata and class has, on the whole, been absent in the study of tribal movements. The reason for this is the assumption that there is no class differentiation in tribal society. However, in more recent writings, this aspect has received attention. Thus, in movements focused on political autonomy, the role of the educated middle class has increasingly been highlighted. Similarly, in movements that were launched in the nineteenth and the early twentieth centuries, the role of religious leaders or leaders who claimed to be incarnations of god has been examined (MacDougall 1985; Singh 1966). The leadership of the movement in general has also been noted. In the early phase of the movement, that is the period 1795–1860, most leaders came from the upper sections of society; in the phase after 1860, most leaders came from the lowest rungs of society (Datta-Ray 1983; Singh 1983; Shah 1982). Yet in all these studies, an an account of the role of women has been conspicuous by its absence. Paradoxically, this is the case even in forest- and agrarian-based movements, which had a greater bearing on the lives of women compared to men. This is not to say that there has been no reference whatsoever of women's role in these movements. Studies of the Tebhaga movement of 1946–47 and the agrarian- and forest-based movements launched by Jharkhand Mukti Morcha (JMM) and Marxist Coordination Committee (MCC) in Jharkhand in the 1970s provide descriptions of women's participation in these movements. Women's participation involved engaging in activities such as forcible harvesting and stacking of crops in their courtyards, keeping watch on trespassers into the village, i.e. the police and the military, attacking the police when they came to arrest leaders, and providing housing, food, and information to members of the underground (Cooper 1988; Singh Roy 1992: 54–60; Sahu 1996: 88). Women's participation was also seen in protest movements against forest contractors and forest officers. Women leaders played a significant role in these activities (Sahu 1996: 108). They have invariably formed a very significant part of various grassroots organizations campaigning for issues such as equal wages, right to work, and campaigns against alcoholism and other social evils. Women's participation was not confined to such movements alone but was also evident in movements such as resistance or rebellion against colonial rule and the demand for autonomy from India or within India. In the case of the former, there have been references to the participation by Santhal women in various resistance movements that were launched by them from the third quarter of the eighteenth century to the first quarter of the twentieth century. In the Santhal rebellion of 1855, of the 25,000 people who had gathered at the Bhagnadihi ground, about half were women and children. Women also marched to Calcutta along with their male partners with a view to placing their grievance before the governor. They suffered military torture and molestation when their men plunged into direct action against the British. Many of them took up native arms and fought against the British until death (Sahu 1996: 105). In addition to their participation in the above struggles, especially the Santhal rebellion and the Birsa movement, the role of tribal women in the peasant and workers' movements of the 1960s and 1970s has also been highlighted (Mitra 1984: 80–90).[4] There have also been some descriptions of women leaders of some movements, such as Rani Gaidinliu and Rajmohini Devi. Gaidinliu, who led an anti-colonial struggle among the Nagas, has been a source of inspiration to thousands of her followers. Similarly, Rajmohini Devi of Sarguja (now in Chhattisgarh) tried, in the early 1950s, to wean the Gond tribesmen away from liquor and magical practices and to impart to them some form of ethical religion. She established an organization called the Bapu Dharm Sabha Adivasi Mandal to this end (Roy Burman 1988: 11–12). However, gender issues in their activities and programmes remain unaddressed in most studies.

Economic and Political Development

In the post-independence period, various provisions have been made by the state for the protection and upliftment of tribes. The benefits under these provisions have been distributed very unevenly across the tribes. Within the tribes too, distribution is very uneven along the line of sex. This is reflected in the negligible presence of tribal women in educational institutions, government and semi-government jobs, and institutions of governance. The female literacy rate among tribes as per the 1991 Census was 18.2 per cent as against 29.6 for men. The data for other sectors along the line of sex are not easily forthcoming. The data that we have for the north-eastern states are quite revealing.

Table 10.1: Percentage Distribution of Workers by Sex

State	Primary sector		Secondary sector		Tertiary sector	
	M	F	M	F	M	F
Arunachal Pradesh	54.6	89.9	12.3	2.2	33.1	7.9
Meghalaya	69.5	83.8	5.0	1.5	25.4	14.7
Mizoram	58.1	78.3	7.2	1.8	34.7	19.9
Nagaland	63.3	92.1	5.1	1.2	31.6	6.7

Source: Women in India: A Statistical Profile, Planning Commission, 2000.

The above data clearly point to the fact that there is a bias against women in the values and social structures of tribal societies, especially in the domain of politics. To ensure the participation of tribes in the decision-making process, seats were reserved for them in Parliament and the state assemblies (lower houses only) for a period of ten years initially, but this has been extended until a date under a constitutional amendment. The participation of tribes in the political process has been studied at various levels, ranging from local self-government bodies, such as panchayats and district councils, to state legislatures and Parliament. Yet the participation of women in these bodies has been negligible, if not altogether absent. Women's membership of state legislatures is a case in point. In Arunachal Pradesh, only one seat was held by a woman out of the total number of 59 seats. The corresponding figures were three out of 57 for Meghalaya, zero out of 40 for Mizoram, and zero out of 60 for Nagaland (Planning Commission 2000).

As mentioned above, women have remained handicapped in almost all sectors of social and economic life. In addition, they have been constrained to work under depressing conditions. They have not only been exposed to greater hardships and difficulties due to depletion of resources (land, forest, water), both at the levels of the household and the community, but also to greater danger due to the changing nature of work and livelihood. Women often walk long distances to collect fuel and to draw water for households or fields. The strategy of coping with such situations and the lack of alternative modes of livelihood have led to large-scale migration of tribal men and women to distant places, even to urban centres, in search of employment. When only men migrate, women have had to bear the burden of taking care of the farm in addition to their usual responsibilities.

The constitutional provisions have affected men and women differently in tribal societies. Nongbri locates these effects within a concrete situation of state action, viz. the Supreme Court ban on timber logging and its implications for the tribal women of the north-eastern region. She points out that this ban has resulted in increasing women's already enormous economic and domestic responsibilities, imposing an additional burden on them. She argues that the ban threatens women's traditional rights to land and forest, and also reinforces their marginalization from the management of their natural productive resources (Nongbri 2001: 1898–1899). Fernandes and Menon express similar views on the impact of the depletion of natural resources on women, who cannot easily find fuelwood, water, fodder, and, in certain cases, even the daily requirements for food (Fernandes and Menon 1987; Menon 1993: 350).

Emerging Discourse

Tribal communities in India are enormously diverse and heterogeneous, differing widely with regard to language, physical characteristics, demographic traits, means of livelihood, and cultural exposure to the wider world. They are, in fact, more heterogeneous than the larger Indian population if caste were excluded from consideration. The People of India project conducted under the auspices of the Anthropological Survey of India enumerated 461 tribal communities, of which 174 have been identified as sub-groups. Hence, the traditional treatment and position of women among different tribes differs, with considerable variations in their roles and activities, rights and privileges. At the same time, women continue to be governed by customary laws and norms. The data that we have on gender in tribal societies are sketchy, especially in the context of recent changes in these societies. Given such heterogeneity, it is difficult to generalise about the position of women as a whole across tribal groups in India. What has been observed can at best be described as being illustrative and heuristic.

Despite such heterogeneity, tribes share one trait in common, that is, they are different from the dominant community of the region, whose members have always been seen as aliens and outsiders. This is seen most clearly in situations where groups identified as tribes are engaged in intense competition and conflict with the dominant regional community, or even with those from outside the dominant community. In north-east India, such conflict resulted not only in the creation of tribal states but also led to an exodus of the non-tribal population from the sub-region. Such conflict still prevails, especially in states or regions where non-tribes still form a significant population. In the rest of India, this is seen most clearly in Jharkhand than in any other region. Outsiders here are invariably described as exploiters and oppressors, and are addressed by terms such as *diku.* In Meghalaya, *dakhar*, a very strong term, is used for outsiders, but carries to a lesser degree the connotations of exploiters and oppressors. In a social arrangement such as this, tribal societies have been experiencing a serious threat to their identity on account of the kinds of changes taking place. There has been a steady erosion of their life-support system based on land and forest. There has been increasing pauperization of the majority of tribes. There is a loss of language. There is a danger of the tribes becoming a minority in their own land. Given all these conditions, the construction of identity and community has been moving in different directions. In all these processes, tradition and customary law, and hence even gender, play a pivotal role in the context of tribal societies.

This has led to an interesting discussion on gender issues in tribal societies with far-reaching consequences for women's freedom, on the one hand, and the issue of equality, on the other hand, especially concerning property rights in land. The problem concerning the above two issues revolves around three points. One concerns tribal tradition. The second relates to transfer of tribal land from tribes to non-tribes. The third concerns the use of the provision of reservation by children born of inter-tribe marriage. It hardly needs to be mentioned that the last has emerged from the encounter of tribes with non-tribes.

One of the most important changes in tribal society is the large-scale alienation of tribal land from tribes to non-tribes. Several studies have pointed to the massive land dispossession that took place during the British period. This process has been accelerated during the post-independence phase of development despite the adoption of the Land Transfer Regulation in states with large tribal populations. Tribal land has passed from tribes to non-tribes in a variety of ways, mostly through fraud and deceit.

One of the ways by which non-tribals acquire tribal land is by marrying tribal women. In view of the fact that there is a restriction on the alienation of land from tribes to non-tribes, such methods have become fairly popular in areas like Jharkhand. Singh refers to the high incidence of alienation of tribal land through marriage with tribal women among the Hos of Singhbhum (Singh 1988: 3). Tribal women entering such marriages are not only seen as aligning with the *dikus* but also as acting as conduits of land transfer from tribes to non-tribes. This is an emotive issue among the tribes in view of the long struggle that they have waged against the alienation of tribal land. Coupled with this is the wider issue of tribal demography. Tribes in their own territory or locality are increasingly shrinking in numbers, with far-reaching economic and political implications. In the process, their survival as a group or community is at stake. Hence, anything that tends to jeopardize their land and population is regarded with a great deal of concern and anxiety. This explains the opposition against women marrying outsiders, especially those considered *diku*.

Related to this concern is the issue of reservation. There has been a general tendency among the families of mixed (inter-tribe) marriages to take advantage of reservation extended for the benefit of tribes. This trend is indeed problematic. Should children born of a tribal mother and a non-tribal father be considered tribal or non-tribal? The question has two aspects. One is legal and the other is socio-cultural. After all, even tribe as a legal category in the form of the scheduled tribe has a socio-cultural basis. Both these aspects are problematic in the case of mixed families. As per the customary law of the community described as a scheduled tribe, lineage is invariably patrilineal, which makes the status of children of mixed marriages problematic. Even if they were to take their mother's totemic title and seek to take advantage of the legal provisions provided for tribes, they would still be contravening customary law. The opposition against women marrying non-tribals is compounded by the weight of tribal tradition, according to which marriage outside the community is regarded a crime as serious as incest.

An equally interesting discourse in tribal society is over the issue of women's property rights, particularly inheritance. A section of tribal society is in favour of property rights in land for women despite resistance to the same in the name of custom and tradition. In Jharkhand, for example, it is argued that as per tribal tradition, there is no individual ownership of land. Rather, the tradition there is that of the *khuntkatti* system, in which land is invariably held on the basis of lineage and hence property belongs to no one individual. Hence the question of extending ownership rights in land to women does not arise. Nevertheless, it is interesting to note Singh's observation on this matter. A higher sex ratio among the Ho tribe accounts for a large number of unmarried women. A number of them claimed their customary right to maintenance, which was often questioned by their male agnates. Singh, as the commissioner of Chhotanagpur, disposed of a large number of cases regarding the right of Ho women to maintenance according to custom (Singh 1988: 3). A similar articulation of identity is also evident in a matrilineal society like the Khasis. However, among the Khasis the issue of ethnic identity is raised along with the demand for changes in the kinship system, viz. the role of kinship and the rules of inheritance (Nongbri 1998, 2000). Resistance to the demand for women's property rights is also related to the economics of land size. It is argued that women's entitlement to land rights will lead to further subdivision and fragmentation of already subdivided and fragmented land in Jharkhand, and hence will have implications for farm efficiency and viability.

Issues such as these pose a problem for a woman as an individual and a citizen, on the one hand, and as a member of a community, on the other hand. As an individual and as a citizen, a woman is entitled to the provision of human rights as well as citizenship, which guarantee the individual the right to freedom. Yet human rights as well as rights under the Indian Constitution also make provision for the protection and safeguarding of tribal interests on the ground of tribes being either a minority or an oppressed social group or community. Adopting either one or the other position can only be taken at the risk of being accused of ignoring or overlooking the other dimension of the problem. This stance is sure to be more problematic for non-tribal scholars or activists than it is for tribals. The problem has to be resolved from within, and none other than by the tribal people themselves.

Notes

1. K. S. Singh has made this observation in the volume brought out by the India Anthropological Socity in 1987.
2. The edited volume by Lucy Zehol contains papers on the status of women in the different Naga tribes.
3. Eleven regional seminars sponsored by ICSSR-NERC were held in different parts of north-east India between May 1998 and February 1989. As many as 136 presentations were made at these seminars. The reports of the directors of the regional seminars were discussed at length at a meeting of social scientists at Guwaheti University, held on 23–25 February 1989. The purpose of this meeting was to distill the information and the various points of view presented by the many papers and discussants.
4. The article of Mitra referred to here is a summarized version of her larger work titled 'Tribal Women in Tribal Movements: A Study of a Particular Movement in Bihar'.

References

Bose, P. K. 1984. *Classes and Class Relations Among Tribes of Bengal.* Delhi: Ajanta Books International.

Cooper, Adrienne. 1988. *Sharecropping and Sharecroppers' Struggles in Bengal, 1930–1950.* Calcutta: K. P. Bagchi & Co.

Chauhan, Abha. 1990. *Tribal Women and Social Change in India.* Etawah: A. C. Brothers.

Elwin, Verrier. 1961. *Nagaland.* Shillong: Research Department, Adviser's Secretariat.

———. 1986. *Baigas.* Delhi: Gyan Publications.

Fernandes, Walter and Geeta Menon. 1987. *Tribal Women and Forest Economy: Deforestation, Exploitation and Social Change.* New Delhi: Indian Social Institute.

Furer-Haimendorf, Christoph von. 1933. *Naked Nagas*. Calcutta: Thacker, Spink & Co.

Ghurye, G. S. 1963. *The Scheduled Tribes*. Bombay: Popular Prakashan.

Gupta, R. K. 1979. 'Witchcraft Murders in the Duars', in J. Troisi (ed.), *The Santhals: Readings in Tribal Life*, Vol. 1*: Religion and Magic*. New Delhi: Indian Social Institute, pp. 265–75.

Hutton, John Henry. 1921. *The Sema Nagas*. Oxford: Oxford University Press.

Kelhou, 1998. 'Women in Angami Society', in L. Zehol (ed.), *Women in Naga Society*. New Delhi: Regency Publications, pp. 51–61.

Kelkar, G. and Dev Nathan. 1993. 'Women's Land Rights and Witches', in M. Miri (ed.), *Continuity and Change in Tribal Society*. Shimla: IIAS, pp. 109–118.

Kishwar, Madhu. 1987. 'Toiling without Rights', *Economic and Political Weekly* 22 (3): 95–101; 22 (4): 149–55; and 22 (5): 194–99.

MacDougall, John. 1985. *Land or Religion: The Sardar and Kherwar Movements in Bihar, 1858–1885*. New Delhi: Manohar.

Mandelbaum, D. G. 1970. *Society in India*. Bombay: Popular Prakashan.

Mann, K. 1987. *Tribal Women in a Changing Society*. Delhi: Mittal Publications.

———. 1988. 'Bride-Price in Tribal North-East India', in J. P. Singh, N. Vyas, and R. S. Manu (eds.), Tribal Women and Development. Jaipur: Rawat Publications, pp. 169–78.

———. 1996. *Tribal Women on the Threshold of Twenty-first Century*. New Delhi: M. D. Publications.

Menon, G. 1993. 'Tribal Women: The Ignored Sector in the Development Debate', in M. Miri (ed.), *Continuity and Change in Tribal Society*. Shimla: IIAS, pp. 350–53.

Misra, P. K. 1984. 'From Bride-Price to Dowry', in K. S. Singh (ed.), *Economies of the Tribes and Their Transformation*. Delhi: Concept Publishing Company.

Mitra, M. 1984. 'Tribal Women and Tribal Movements', *Samya Shakti—A Journal of Women's Study* 1 (2): 80–90.

Nathan, Dev. 1997. 'Gender Transformations in Tribes', in Dev Nathan (ed.), *From Tribe to Caste*. Shimla: IIAS, pp. 247–86.

Nongbri, T. 1998. 'Gender Issues and Tribal Development', in RGICS Paper No. 47 (Problems in Tribal Society: Some Aspects).

———. 2000. 'Khasi Women and Matriliny: Transformations in Gender Relations.' *Gender, Technology and Development* 4 (3): 359–95.

———. 2001. 'Timber Ban in North-East India: Effects on Livelihood and Gender.' *Economic and Political Weekly* 36 (21): 1893–1900.

Pathy, J. 1984. *Tribal Peasantry: Dynamics of Development*. Delhi: Inter-India Publications.

Punalekar, S. P. 1988. 'Tribal Women and Development Milieu: Social Context and Emerging Contradictions', in J. P. Singh et al (eds), Tribal Women and Development. Jaipur: Rawat Publications.

Roy Burman, B. K. 1994. *Tribes in Perspective*. Delhi: Mittal Publications.

———. 1988. 'Challenges of Development and Tribal Women', in J. P. Singh (ed.), *Tribal Women and Development*. Jaipur: Rawat Publications.

Sahu, Chaturbhuj. 1996. *The Santhal Women: A Social Profile*. Delhi: Sarup & Sons.

Shah, Ghanshyam. 1982. 'Tribal Issues: Problems and Perspectives', in B. Chaudhuri (ed.), *Tribal Transformation in India*, Vol. 2, Delhi: Inter-India Publications, pp. 113–41.

———. 1990. *Social Movements in India. A Review of the Literature*. Delhi: Sage Publications.

Singha Roy, Debal K. 1992. *Women in Naxalite Movement: Tebhaga, Naxalite and After*. New Delhi: Manohar.

Singh, K. S. 1983. *Tribal Movements in India*. Vol. 2. New Delhi: Manohar.

———. 1966. *The Dust-Storm and the Hanging Mist: A Study of Birsa Munda and His Movement in Chhotanagpur*. Calcutta: Firma K. L. Mukhopadhyay.

———. 1988. 'Tribal Women: An Anthropological Perspective', in J. P. Singh (ed.), *Tribal Women and Development*. Jaipur: Rawat Publications, pp. 3–10.

Sachchidananda. 1988. 'Tribal Women in the Throes of Development', in J. P. Singh (ed.), *Tribal Women and Development*. Jaipur: Rawat Publications, pp. 79–90.

Sinha, S. 1958. 'Tribal Culture of Peninsular India as a Dimension of Little Tradition in the Study of Indian Civilization: A Preliminary Statement', *Journal of American Folklore* 71(July–Sept): 504–518.

Unnithan-Kumar, Maya. 1997. *Identity, Gender and Poverty: New Perspectives on Caste and Tribe in Rajasthan*. Oxford: Berghahn Books.

Vidyarthi, L. P. 1982. 'Research on Tribal Culture of India', in P. K. B. Nayar (ed.), *Sociology in India: Retrospect and Prospect*. Delhi: B. R. Publishing Corporation.

Zehol, L. (ed.). 1988a. *Women in Naga Society*. New Delhi: Regency Publications.

Zehol, L. 1988b. 'The Tangkhul Women Today', in L. Zehol (ed.), *Women in Naga Society*. New Delhi: Regency Publications, pp. 20–29.

Index

About the Author

Virginius Xaxa is Professor of sociology at the Department of Sociology, Delhi School of Economics, University of Delhi, and Rajiv Gandhi Chair in Contemporary Studies, North-Eastern Hill University, Shillong. He obtained his Ph. D. from the Indian Institute of Technology, Kanpur. He was a Post-Doctoral Fellow at the Ecole des Hautes Etudes en Science Sociale, Paris, the recipient of Commonwealth Academic Staff Fellowship, SOAS, University of London, and a Fulbright Visiting Lecturer at the University of California, Santa Cruz, USA. Besides numerous papers in national and international journals, he has published two books: *Tea Plantation Labour in India* (with Sharit K. Bhowmik and M.A. Kalam, 1996) and *Economic Dualism and Structure of Class: A Study of Plantation and Peasant Settings in North Bengal (1997)*. His research interests are in the areas of rural sociology, tribal studies, and ethnicity and religious identity.